1977

$2.00
2-24-21

Environment

Also by John G. Fuller

THE POISON THAT FELL FROM THE SKY

JOHN G. FULLER

THE POISON THAT FELL FROM THE SKY

RANDOM HOUSE · NEW YORK

All rights reserved under International and Pan-American
Copyright Conventions. Published in the United States by
Random House, Inc., New York, and simultaneously in Canada by
Random House of Canada Limited, Toronto.

Grateful acknowledgment is made to the following:
Houghton Mifflin Company: For an adaptation of Chapter 1
from *Silent Spring* by Rachel Carson. Reprinted by permission.

Library of Congress Cataloging in Publication Data
Fuller, John Grant, 1913–
The poison that fell from the sky.
1. Chemical industries—Italy—Meda—Accidents.
2. Icmesa. 3. Air—Pollution—Italy—Seveso.
4. Tetrachlorodibenzodioxin—Toxicology—Italy—Seveso.
I. Title. RA578.C5F84 945′.21 77–13540
ISBN 0–394–42495–6

Manufactured in the United States of America
2 4 6 8 9 7 5 3
First Edition

FROM SILENT SPRING
BY RACHEL CARSON
(1962)

There was once a town where all life seemed to live in harmony with its surroundings. . . .

Then a strange blight crept over the area. . . . Mysterious maladies swept the flocks of chickens; the cattle and sheep sickened and died. Everywhere was the shadow of death. The farmers spoke of much illness in their families. In the town, the doctors had become more and more puzzled by new kinds of sickness. . . .

In the gutters under the eaves and between the shingles of the roofs, a white granular powder still showed a few patches; some weeks before, it had fallen like snow upon the roofs and the lawns, the fields and streams. . . .

This town does not actually exist. . . . I know of no community that has experienced all the misfortunes I describe. . . . (Italics added)

FOREWORD

THE STORY that follows happened to take place near Milan, Italy. It could easily have happened in Connecticut, Ohio, Virginia, Michigan—or any other location where the $100-billion-a-year chemical industry is churning out over 30,000 different chemicals, most of them untested for safety.

More often than not, these chemicals leak out slowly and insidiously, as in the Kepone incident in Virginia or the PBB invasion in Michigan. But the tragedy near Milan was sudden, direct, concentrated, and awesome. It symbolizes the era of the chemical plague just as surely as Hiroshima signaled the tragedy of the atomic age.

In 1962, Rachel Carson opened her book with a story similar to this one, but the incident she described was written as a myth, a legend. Critics were quick to seize on this, to point out that all of *Silent Spring* was merely a legend, and therefore invalid.

Regrettably, the story that follows has changed the myth to a tragic reality, fourteen years after *Silent Spring*. It has created a medical and scientific earthquake that is still rumbling throughout the world, and will undoubtedly continue long into the future.

J.G.F.

THE
POISON
THAT
FELL
FROM
THE
SKY

1

THE AUTOSTRADA from Milan to Como strikes almost due
north toward the Alps and the northern Italian lakes that have
been sung about since the beginning of history. Halfway between, on the soft green carpet of the Po Valley, a sign points
to the town of Seveso. The ancient buildings there are now
crowded by modern factories, part of the industrial tentacles
of Milan that have reached out to grapple them. But there are
still the wide fields of wheat and corn, the verdant orchards,
the long rows of mulberry bushes, and the villas built by the
artisans themselves with loving care.

Such was the home of Gino Zorzi, built with enormous pride
over a twenty-year span on weekends and holidays, mostly on
his time off from his job as a building contractor. There was no
homemade look about it: handsome parquet floors, recessed
lighting, thick masonry walls against the heat of the summer
and the cold wind from the Alps in the winter, a sturdy red-tile
roof, and modern architectural design. There was an electri-
cally operated front gate, with phone communication to the
house. Around it along the Via Carlo Porta were other houses
like it, interspersed with lush green fields, built in the same
way and with the same care.

The summer of 1976 had been kind to the community of Seveso, and for the miniature resident-farms of the people living on the Via Carlo Porta. Seveso had been spared the ravages of the drought in northern Europe that summer. Gino Zorzi was proud of his vegetable garden, as well as his flowers. There was a bountiful crop to be canned or frozen. As a family man, Gino was delighted with the nearby town recreation center, with its baseball field and outdoor swimming pool. It was a boon for his children, Fabio, eight, and Giuliana, eighteen.

Saturday, July 10, 1976, was a special day for all the Zorzis. Giuliana was about to celebrate her eighteenth birthday with a gala party on the Zorzi terrace garden that evening. When an Italian girl reaches eighteen, it is always a joyous event. At noon, there was already excitement in the air. Milena, her mother, was out on the terrace, putting the finishing touches to the table decorations with motherly care and attention. Although Giuliana wanted to help, her mother firmly shooed her off to the beauty parlor so that she would look perfect when her fiancé arrived that evening.

Content with the table decorations, Milena Zorzi went into the house to prepare it for the guests. She was a meticulous housewife, proud of her furnishings and the gracious atmosphere she had created in her home. At half past twelve, she decided to take a break and have lunch with her husband in the kitchen.

Then, within moments, it happened: A loud, screeching, hissing sound. It was ugly and ominous. They had never heard anything quite like it before.

A few houses down on the Via Carlo Porta, Mrs. Licio Cassio was just returning from shopping for her family. With her were her two younger children, Elda, twelve, and Paola, ten.

They were glad to return home, because after lunch they planned to run over to the nearby community swimming pool to join their friends in the warm Italian sun.

Carlo, the twenty-year-old son, was in his room. He had just returned from the library with some books on architecture for his course at a local university. Beside him were Lilli and Fanny, his two Siamese cats he was so fond of. Carlo had already put a lot of his skills to work in helping his father build the Cassio home. Like the neighboring Zorzis, the Cassios built their home over a period of time, some of it reflecting Carlo's early-blossoming talent. His bedroom was of his own design, built like an architect's studio, with a separate entrance. Among the neighbors, there was a touch of friendly envy about the Cassios' beautiful marble entrance and its varicolored flagstone terrace.

As with the other houses in the neighborhood, the Cassio home also served as a small farm. There were rabbits outside in their cages, quietly munching enormous quantities of grass collected from the nearby fields. The rabbits were a particular delicacy. Rabbit stew or ragout were special favorites. They were raised continuously, and later stored in the ample modern freezers found in most of the homes in the area.

To make the family budget stretch, the food they grew in their garden, and the rabbits, chickens, and goats were invaluable. The fruit trees, too, laden with peaches, plums, apricots, and pears aided the family budget.

At the moment that Mrs. Cassio put down her groceries on that Saturday in July, she was startled by the same terrifying whistling noise. On the wall, the clock read 12:38.

At a little before twelve-thirty that afternoon Viro Romani, a technician at the Icmesa chemical factory, was finishing up his lunch in the plant cafeteria. There was no active production going on at the factory that Saturday. The last shift had closed down at six in the morning. Only about 10 of the 160 employees were on hand for maintenance and clean-up. In addition to essences and essential oils for perfumes and cosmetics, the Icmesa plant manufactured trichlorophenol, a raw chemical used for making a weed-killer and defoliant called 2,4,5-T, and also hexachlorophene, a bactericide.

Icmesa made neither of these as end products. The trichlorophenol—known as TCP—was made exclusively for the Swiss parent company, Givaudan. This company in turn was a subsidiary of Hoffmann–La Roche, one of the largest pharmaceutical companies in the world. Givaudan used the TCP only for making hexachlorophene, the principal ingredient in many surgical soaps. After a long, toxic, unhealthsome track record, hexachlorophene had been banned from wide use in toiletries in the United States. But it was still available in other countries with less stringent regulations.

The chemical reactor for making TCP had been shut down that Saturday morning with the ending of the night shift. Viro did not work in that section; his maintenance duties took him to another part of the plant. After lunch, he came up from the cafeteria to the traditional coffee bar with a few other workers for a cup of espresso.

Before he had a chance to sip it, he was startled by an inordinately loud *pong*—a noise that made all of the workers at the bar jump. The sound was immediately followed by a deafening whistling noise.

Viro and his companions ran outside and looked up at the sky. A huge, grayish-white cloud was spewing out under tremendous, screeching pressure from the safety-valve stack of the TCP reactor. In moments, minute particles like very fine sand or dust were falling around them. They covered Viro's face. It felt like half-wet salt. He wiped his face with a maintenance rag, his skin smarting. A thick white fog surrounded all of them. The leaves and the ground and the factory roof were immediately covered with tiny white crystals.

Running back inside the factory, he saw Gallanti, the foreman, rush into the reactor room, a gas mask on his face. Gallanti lurched toward the reactor and quickly turned a metal valve wheel to let water rush into the reactor's cooling system, damping the tremendous heat that had obviously built up. But more than five minutes had gone by. The enormous cloud, billowing out in a thick mass, was rolling slowly but inexorably to the south, over part of Seveso, toward Milan.

An overpowering, medicinal and chlorine-like stench permeated everywhere, choking them all. Within fifteen minutes, the head of the laboratory arrived. He instructed all the men to go home and take showers. In the light wind, the cloud, in a typical Gaussian plume, hung almost motionless toward the southeast. By now it stretched some five miles long, with the small white crystals dropping silently like snow over the rooftops and fields in its path. No one knew at the time that the cloud had been inadvertently saturated with one of the deadliest poisons known to mankind.

At the moment they first heard the awesome sound, Gino and Milena Zorzi rushed out to the terrace of their home. They glanced to the north, in the direction of Lake Como and the Alps. But there was no need to look that far. Just beyond the neighboring houses, not more than a mile away, the enormous cloud was forming, filling the sky and tumbling toward them like a giant ice cream cone. It was thick and gray, rolling over on itself, then suddenly changing to several different colors. In moments it was over their heads, thick and lazy in the light wind, gliding southward toward Milan.

Out of the cloud, the mist began to fall. With it came the

stench, vicious and acrid. The fog was settling down every-where, on the trees, the grass, the cornfield across the way—and to Milena's great distress, on the table decorations. Clamping a handkerchief over her face, Milena ran out to the terrace garden to take in the plates from the tables.

"What an awful thing to happen on Giuliana's birthday!" she called to her husband.

Coughing and rubbing his eyes, which had suddenly begun to burn and sting, Gino Zorzi ran to close the windows and the rolling shutters in the house.

The reaction in the Cassio house was the same. Mrs. Cassio and her children ran outside to watch the giant cloud approach. Without knowing why, she felt that some kind of a gas tank had exploded. Then, when the damp crystals of the fog descended on her, she rushed the children back into the house and quickly rolled down the shutters. As the fumes began to choke them, they clamped handkerchiefs over their mouths.

Alarmed, Carlo left his drawing board and ran to the house of a nearby friend. There, the grandmother, with a weak heart, had fainted. Carlo and his friend were able to revive her; then he called the local police.

The questions were many: What had happened? Where did the cloud come from? There had always been the smoke from the factories that ring Milan and its suburbs—but this was different, suffocating, threatening.

The local police were unable to clarify the situation. They were in the process of investigating it, they said, because there had already been many complaints. They recommended that Carlo call the carabinieri—the national police. They might know more.

Carlo did so. But there was still little information: only that

8

the cloud had come from an accident in the Icmesa factory, just over the Seveso border, in Meda.

Carlo hung up. He told his friend the little he was able to find out. Neither of them was satisfied. They left to go to the factory.

When they arrived, they were met by a group of technicians. They were told there was nothing to worry about—it was just a slight mishap, and everything would be normal when the air cleared. There was no further information forthcoming. Still puzzled and curious, Carlo and his friend made their way back to their homes on the Via Carlo Porta, less than a mile away, still feeling the sting from the cloud that hung over the area.

Back on their street, they found the neighbors complaining of headaches and burning skin and eyes. But this was not uncommon, in a much milder form. The persistent effluence from the many factories scattered around the area was always with them. Many grumbled, and some talked of lodging a major complaint. Knowing that this had been ineffectual in the past, the neighbors decided to wait until they learned more from the authorities.

Meanwhile, Milena Zorzi was intent on cleaning up the garden terrace for Giuliana's birthday party; she was determined that nothing would stop that. She carefully wiped the tables, cleaning off the white dust that had settled like brine on them. Then she picked some choice plums, peaches, and pears, wiped them carefully, and put them in bowls as centerpieces for the tables.

The party on the terrace was lively. It continued until the small hours of the morning. But the merriment was somewhat forced. The guests for the most part tried to forget the acrid smell and the cloud that still hung over the houses and the fields.

In the morning, the Zorzis had much to do. After cleaning up the terrace, they had to catch the morning bus to the lake region, where Fabio, their eight-year-old son, was in camp. Waiting for the bus, the Zorzis noticed that the white crystals of the day before had changed in texture. All over the trees, the plants, the fields, and the gardens, a shiny, oily coating now glistened in the morning sun. Nothing seemed to have escaped it. Though not as thick as it once was, the cloud still lingered above, floating in the stillness of the bright July morning.

At the Cassio house down the street, Mrs. Cassio woke up with a throbbing headache and swollen eyes. The chemical smell was so strong that she felt on the edge of vomiting. If the symptoms persisted, she thought, she would go to the doctor. Meanwhile, she rose and dressed and went to help her husband feed the rabbits and gather some vegetables for the ritual of their special Sunday meal.

In spite of the shiny coating, they picked the choice ripe tomatoes, the corn, the green beans and peppers. They tried to put the ominous cloud hanging overhead out of their minds. Their street, their house, their garden still seemed bright and beautiful, in spite of the oily film that covered the greenery. Mrs. Cassio felt confident that her headache and swollen eyes would disappear the next day.

AT HIS HOME in Seveso on Sunday, July 11, Mayor Francesco Rocca was facing complaints of the cloud and the smell and the irritation, and was waiting for the report of his local health officers, who had just begun an investigation of the matter. When they arrived, all they were able to report was that "something like a defoliant" had escaped from the Icmesa factory. A full report was to be given to them Monday, when the factory reopened. Meanwhile, the health officers told the mayor that they suspected that the product was TCP, a severe irritant, but not expected to be seriously damaging.

Not long after they left, two more visitors arrived at the mayor's house, technicians from the Icmesa plant. They were unable to add to what the health officers had already reported. The chemicals had escaped in considerable quantity, and samples had been sent to Switzerland for the exact analysis. They suggested that just as a precautionary measure it might be best to notify the people in the Seveso area not to eat the fruits and vegetables from their gardens for a while, pending identification of the exact composition of the product. They did not seem to be too concerned.

Mayor Rocca pressed for more details, but there was little

more they could tell him. Following their suggestion, however, the mayor issued a statement for the press and radio about the fruits and vegetables, indicating that more information would follow. He felt frustrated, but knew he could only await with interest the report of his health officers after they visited the plant the next day.

As the mayor pondered the strange situation, the children of Via Carlo Porta and the neighboring streets were swimming, playing at the recreation center, and flocking to the baseball diamond, undaunted by the waxy coating that seemed to cover everything exposed to the sky.

It was a warm, sunny day, and the pool was inviting in spite of its medicinal smell. On the baseball diamond, a lusty game was in progress as the younger kids cheered and romped and rolled on the grass. The peaches, pears, and plums were especially tempting, and other children were gathering the fruit, eating some of it, and taking some home.

In some of the yards of the many miniature farms in Seveso, chickens and rabbits were selected for the Sunday evening meal, along with the fruits and vegetables already gathered. As the children came in from their holiday playing, they were scrubbed and readied for dinner, cooked with typical Italian flair.

At the Cassio home, the Sunday dinner was marred by Mrs. Cassio's headache and swollen eyes. They had grown markedly worse, and she was forced to leave the table early. She was not alone; the others in the family were affected in one way or another. But her reaction was more severe. As she kissed her family good night, she resolved to see her family doctor in the morning.

At the Icmesa plant the next morning, Monday, July 12, all was apparently normal again. Only the immediate area of the

12

TCP reactor was roped off. Not too much was officially said about it. The smell, however, was still overpowering. The workers knew their jobs, but not much about the theory of TCP and its complex reactions.

They talked among themselves, however, about what might have happened. The reactor temperature, they knew, usually ran between 140 and 170 degrees C. The safety valve would not have blown unless the temperature had gone over 240 degrees, indicating a violent and unexpected heat rise. The wind that had come up Sunday evening had mostly dispersed the lingering cloud, a violent wind that often swept down from the Alps. With it, part of the fallout had been scattered beyond the area in which it had originally settled.

When the local health officials arrived, they noted in a cursory examination that the workers did not seem to be seriously affected, in spite of some mild headaches, slight nausea, and burning sensations on the skin. In the light of this, they brought back a relatively reassuring report to the mayor. The tests being completed in Switzerland would take an indeterminate time, and any further action would have to wait for the results. The mayor's press release was given short shrift—a small story on the inside pages of the local paper, with a condensed version in the Milan papers without editorial comment.

In spite of reassurances, the workers were somewhat suspicious. They went about wiping up the fine white dust that had settled on all the flat surfaces of the factory, the tables and benches and floors. They were not too happy with the management of Icmesa or its Swiss parent companies. Some of them were flatly accusing the Swiss of using Italy as a sewer, charging that the Swiss companies were acquiring Italian subsidiaries because the pollution laws in Italy were either nonexistent or lacking in enforcement.

Outside the factory, the Icmesa technicians were moving out farther from the factory grounds to get more samples of leaves and grass and soil to ship to Switzerland for more extensive tests. They worked quietly and inconspicuously, and were hardly noticed.

At her home on Via Carlo Porta, Mrs. Cassio was definitely not improving. On Monday morning she went to the family doctor. Her abdomen was swollen as well as her eyes; her back was in pain; her headache was still relentless. About all her doctor was able to tell her was that she must have had an acute allergic reaction to something. Whether it was the cloud or not, he was unable to say.

He gave her some pills, but she was wary of taking them. Her vague anxiety was increasing. An intelligent and perceptive woman, she realized that inhaling the particles from the cloud was bound to have an effect on all her family. But it was not until the next morning, Tuesday, when she found her fears confirmed. Her youngest daughter's face was covered with a thick, ugly rash.

By Wednesday, July 14, four days after the cloud had spread across the Po Valley, the local doctors were being inundated with similar complaints. They began to compare notes. The symptoms everywhere were very much the same as those experienced by the Zorzis and the Cassios. There seemed to be little question that the illness sprang from the cloud. But there was little the doctors could do without knowing specifically what it was that had contaminated the area, and what antidote there was for it. No news had yet arrived from Switzerland, and the Icmesa officials claimed they still knew nothing beyond the fact that TCP was a strong irritant. While the symptoms were mild, they were so widespread that there was growing concern.

The Icmesa technicians were now moving still farther out from the factory, continuing to collect soil and grass samples. When they were asked informally about the situation, they replied that there was nothing to worry about; there was no danger; everything would be under control. But the uneasiness among the 17,000 residents of Seveso continued. All through the area, the ugly rashes covering the legs, faces, and arms of the children became worse. But there was more to contend with. On Wednesday, many of the local children began to vomit.

Before dinner on Wednesday, Licio Cassio, in the living room of his home on Via Carlo Porta, was sitting with the door open, looking out over the balcony and worrying about his wife and children. Mrs. Cassio had been in bed now for the last two days. She was not improving, nor were the children. The anxiety and uncertainty were becoming almost unbearable. Outside, beyond the balcony, Mr. Cassio noted that the land and house looked very much the same, although some of the leaves were turning a strange yellow. The sky was blue, with the evening sun striking the trees in the background with a pleasing roseate glow.

A lone robin was outside, but something about the way it was flying caught Mr. Cassio's eye. The bird seemed to be moving in an erratic path toward his balcony. He was startled because it was coming directly toward him. In a matter of seconds the robin drifted like a falling leaf, over the balcony railing, through the open door and into the living room. Then suddenly it crashed down on the floor beside him and lay motionless.

Mr. Cassio got out of the chair and went over to the bird. It was barely breathing. He picked it up, but it did not flutter its wings. He took it out to the balcony and tried to revive it with

15

the warmth of his hands. It seemed to respond. Then he lifted it over the rail with both hands, hoping that the bird would fly again. But it didn't. It plummeted straight to the ground. Feeling vaguely uneasy, Mr. Cassio left the balcony and went back in the house.

Later that evening young Carlo Cassio, his two Siamese cats with him, was studying his architectural books in his room. He noticed that Lilli was not stretching with her usual ease. When she tried to, she toppled over on her side. Fanny, the other cat, was encountering a similar problem. She was walking across the floor with a strange, swaying motion, almost as if she were drunk. Both were beginning to cry, in a deep, throaty tone alien to them. He decided that if they didn't improve, he would take them to the vet in the morning. He did not want to voice his fears: were the birds poisoned, and did the cats eat them?

But this was only one of an increasing number of incidents involving animals throughout the neighborhood. Mr. Zorzi, on his evening walk, was startled to see a friend's dog wobbling and staggering toward him, behavior far out of character from the dog's usual liveliness. Almost immediately afterward, he noticed several birds flying in a bizarre manner overhead. Two of them seemed to falter, then both plunged to the ground some distance from him. When he returned home, he was too uncomfortable to mention the incidents to his wife.

BY WEDNESDAY EVENING, Mayor Rocca had been besieged by inquiries, but had no specific information to answer them. The anxiety of the doctors and families was all concentrated on his office. Resentment among the workers at the Icmesa factory was growing; they felt strongly that their supervisors were not giving them the full story. The usual labor-management meeting on Wednesday had been canceled without explanation; the workers had been warned to take showers before leaving work that day, with no reason given.

A phone call from Switzerland Wednesday night, however, brought the day to a climax for the mayor. The news was very disturbing: the initial tests had shown that there were trace amounts of a "very poisonous substance" in the crystals that the cloud had dispersed across the countryside.

But the tests were incomplete. It was important not to be alarmist about the situation. Panic in the town would only make matters worse. At a meeting the next day with the mayors of both Seveso and Meda, all the Icmesa officials could say was that they were pressing Switzerland for the exact nature and amount of the poison. There was no question, they added, that more protective measures were now necessary, and the warnings should be fortified with posters and signs.

At noon on Thursday, Mrs. Zorzi went out to the yard to select two chickens for dinner. With a start she noticed the condition of the tomato plants. They seemed to be burned brown, the leaves dry and crumbly. She passed by the rabbit cage and stopped to look into it.

The first thing she noticed was that they had not been eating. She looked closer. She was shocked to see that blood was oozing from their mouths and rectums. She half-ran to the chicken roost and opened the hatch. All of them had toppled over on their sides and were dead. A thought quickly flashed through her mind: she and her family had been eating their livestock and vegetables ever since the cloud had come over, five days before. She froze with fear.

Because the entire situation had not attracted any real press or radio attention, the news about the strange set of circumstances traveled slowly. The emergency warning in Seveso was still barely noticed beyond the borders of the factory. The incidents involving strangely behaving or dying birds and animals were scattered. The doctors were more annoyed at the lack of information than fearful. There was widespread resentment at the factory, but there was also a certain amount of inertia and apathy. The word-of-mouth stories going around were discounted as rumors, and the warnings. from the mayor's office were regarded as applying only to the area around the factory.

It was mainly the workers at the plant who were growing more restless and impatient. They had learned that Icmesa technicians had quietly been collecting dead animals from the area, and that many more birds had been falling from the sky than anyone had suspected. They sensed that much more information was being kept from them.

On Thursday afternoon at two o'clock, the factory director called a meeting. It included the heads of all the various sections and the head of the union. The atmosphere was tense.

Without explaining why, the director stated that all workmen without exception were to take showers in the factory, and to leave all of their work clothes there. They would be permitted to wear their street clothes home, but their work clothes would be burned. They were told to remain calm, for according to the director there was nothing really dangerous about the situation. The burning of the clothes was simply a precaution. They were instructed to come to work the next day, Friday, as usual.

They did, but discontent with the situation was growing strongly. As soon as they arrived on Friday, thirty workers were assigned to a special detail. They were given tall posts and instructed to place them in designated areas around the factory; tacked to the posts were signs reading DANGER AREA. CONTAMINATED. DO NOT EAT VEGETABLES, FRUIT, OR ANIMALS THAT EAT GRASS FROM THE GROUND.

The last phrase was something entirely new. It was ominous. All along, even those who'd been refraining from eating the fruits and vegetables had been eating the rabbits, chickens, and lambs they'd raised themselves. Rabbits, the favorites, were enormous grass-eaters. Each day the residents collected grass from the nearby lawns and fields, and put it in the cages for the rabbits to eat.

In one sense, the posting of notices in the immediate area of the factory was reassuring. The area posted was not large, and those who lived outside it felt a sense of relief. But the workers were not easily mollified. By noon on Friday the union committee asked for a meeting with the factory director. He refused, saying that time pressures did not permit him to join them.

19

Action by the union committee was immediate. They called a meeting of all the workers on the day shift, over a hundred strong, without the director's consent. When they gathered in an open plant area, the director appeared at the meeting, joined by the factory supervisory staff. The workers balked, but the management group remained. After nearly ten minutes of haggling, Roberto Chiappini, the fiery young leader of the union, stated flatly that unless the management group left the meeting within five minutes, he would stop it and hold it outside the factory. The managers refused, and the workers filed out of the building.

On the street, it did not take long for the workers to agree that the situation was much more serious than they had been led to believe. They had been kept in the dark about a situation which affected their lives. It seemed obvious to them that they were not being told the whole story. Passions were running high. Although they had been told to report to work on the following Monday, they unanimously voted to strike. Then, grimly, they voted to assemble in front of the municipal hall both in Seveso and in Meda, where the plant was actually located, while the union leaders demanded that the factory be closed until safety could be assured.

When they gathered in front of the Municipal Hall in Seveso, the police were startled. Seveso had been noted as a quiet town. The people here had always been happy and tranquil, hard-working artisans who had been laboring over the years to raise their standard of living to equal the highest in Italy. The police had never had to deal with a problem in crowd control, but they did their best as the union leaders finally entered Mayor Rocca's office.

The mayor, they found, was in total agreement with them. If the warnings had to be increased to such an extent, the factory should not be operating. He assured the workers that he and the mayor of Meda, who also agreed with them, would take immediate legal action. The Icmesa officials responded with a statement that the employees were simply work-shy and were looking for an excuse to "lay down their tools."

The press immediately picked this up, much to the consternation of Icmesa. Very clear battle lines were forming, with Icmesa and its Swiss parent companies on the one side and the workers and local authorities on the other. Meanwhile, more animals were dying. Veterinarians were so busy they had no time to coordinate information.

On Friday evening, a two-year-old baby was rushed to the hospital with large, running sores all over his body. It was the first hospital case arising from the poison cloud of nearly a week before.

IV

ON SATURDAY MORNING, smack in the middle of the unwanted crisis, Mayor Rocca awaited the arrival of a Swiss scientist sent by Hoffmann–La Roche and Givaudan who was supposed to clarify the recent news about the additional poison discovered in the animal, soil, and plant samples. The mayor waited with a sense of apprehension and relief. Good or bad, he wanted to know. He was also awaiting the decision of the regional judge in the neighboring town of Desio concerning his demand for the closing down of the Icmesa factory.

Another visitor scheduled for the morning was Professor Aldo Cavallaro of the Provincial Laboratory of Hygiene in Milan. Cavallaro had become convinced that the recent events dramatically illustrated that the disaster was growing and was regional in scope, if not national. He had sensed there was much more to the picture than had been admitted by Icmesa. He had already been out in the fields and orchards, gathering leaves and plants with his bare hands to have them analyzed at the Milan laboratories. He would then compare these results with those of the Swiss scientists who had been so slow in reporting their findings. At this point a full week had gone by, with no really clear information about the composition of the cloud.

Meanwhile, more ominous news was coming in. A dozen more children were being sent to the hospital with massive sores all over their bodies. Several adults had also been admitted with nausea, vomiting, acne, and severe liver and kidney pains. Among these was Mrs. Cassio, whose condition had been growing markedly worse over the week.

By Saturday, the local population did not need a scientific explanation to see what was happening. Birds were now dropping from the sky en masse. Their bodies were lying everywhere. Pet dogs and cats continued to sway drunkenly in the streets; some died. Many rabbits and chickens had already hemorrhaged and died in the yards. Lilli and Fanny, Carlo Cassio's two Siamese cats, continued to be in agonizing pain. Their yowling was impossible to bear. Carlo took them to the veterinarian's office, trying in vain to comfort them.

Dr. Cigognetti, a young Seveso vet, greeted him solemnly. His office had been overrun, leaving him no time to make a coherent analysis of what was happening, although the wave of animal deaths was appallingly evident. Medication was ineffective; the killer was still a mystery.

He examined Lilli and Fanny. His verdict was almost immediate. They were dying. Carlo agreed that they should be put out of their misery. He watched in sorrow as the vet injected them with medication and they quietly died.

Meanwhile Chief Alfonso Calo, head of the Seveso police, had just returned from his vacation to find the most incredible scene in his twenty-two years of service. The contrast with the comfortable scenes of his holiday was appalling. He saw children covered with lesions, animals dying before his eyes, journalists swarming over the quiet streets of the town, and people everywhere frightened and confused. So many people were coming down with rashes that one of his first jobs was to help set up first-aid posts in different sections of the town. One of the greatest fears, he found, was that people were beginning to believe that a plague was striking and spreading swiftly. The dead birds lying in the fields were particularly frightening, along with the swollen faces of children, many of which had ugly pustules and running sores. Passing by the factory in the squad car, he saw the warning signs, and next to one of them, the bountiful peach tree, always so popular with the workers. All the fruit from the lower branches had been picked. His fervent hope was that whatever the contamination was, it would be confined to the small area near the factory.

The concern was centered in Seveso, immediately downwind from the Icmesa safety valve that had spewed out the TCP and the undefined poisons that accompanied it. Residents of the surrounding towns considered themselves lucky to have been at a greater distance from the factory. In Cesano Maderno, two miles south of Icmesa by way of Route N-35, most people felt that the stories filtering in from Seveso were vague and unsubstantiated.

On Tuesday morning, July 20, Dr. Ciccardi, the chief veterinarian of Cesano Maderno, returned from his vacation. He had heard nothing yet from his colleagues in Seveso. He felt sure that there must have been a great deal of exaggeration going around. He was not very far along in his morning calls, however, when he encountered a friend, a truck driver who lived considerably south of the Seveso border.

"Do you know," the truck driver told him, "that all the ducks and rabbits in my courtyard have suddenly died?"

After promising to investigate the matter immediately, Dr.

Ciccardi lost no time in phoning his counterpart in Seveso, Dr. Cigognetti.

"Yes," he was told. "We've found dead animals all the way down to your town line. The situation looks bad."

The shaky comfort that the Cesano Maderno residents had felt was shattered. The contamination had crept silently over their town line. The strange deaths were spreading and no one seemed to know where they would stop.

In Milan, reporters were belatedly realizing that the poison cloud might be a story of major significance. One of them was a television newsman named Bruno Ambrosi. He had some knowledge of chemistry, and after studying the rather scant news reports, he suspected that this might not be an ordinary type of industrial accident. His schedule was already jammed, but the answers he got to some preliminary inquiries disturbed him greatly. A public inquiry in Desio on the Sunday a week after the accident produced only a confused and ambiguous picture. He was unable to find out what unannounced impurities had accompanied the TCP. The Swiss companies stated that it would be at least eight to ten *more* days before their tests would be complete enough to determine the lines of demarcation between the contaminated and uncontaminated zones. The main question that remained unanswered was: Contaminated with what? Whatever it was, it had to be powerful. TCP had no such track record.

On Tuesday, July 20, ten days after the accident, Ambrosi and his TV 2 camera crew from Milan arrived in Seveso. Ambrosi was more than just a former chemistry student; he was also an investigative reporter who liked to dig deeply under the surface. He knew vaguely of TCP from his background, and that it was related to the defoliant 2,4,5-T, which had created much devastation in Vietnam. In Vietnam, the defoliant had been reported to have contributed to an abnormal increase in cancer of the liver, as well as birth abnormalities. His recollection was sketchy, but Ambrosi was determined to look into this further.

Interviewing the police chief in Seveso, Ambrosi found that everyone seemed equally unsure about the composition of the cloud. He was surprised to discover that no one had consulted the doctors at the internationally known Mario Negri Institute, in Milan, just ten miles away from Seveso. This was a well-known nonprofit biomedical-research foundation with sophisticated equipment and a staff of high-ranking scientists. Ambrosi had just covered a major international symposium on pharmacology at the institute the previous week. If anybody could make sense of the confusing reports from Switzerland, this institute should.

Ambrosi used the police chief's phone to call Dr. Luciano Manara, chief of the institute's drug-metabolism laboratory. Since the medical library at the institute was among the best in the world, Manara was sure he could quickly assemble some information on the nature of TCP and its impurities. He told Ambrosi he would call him back in about ten minutes, after checking. Ambrosi's next interview was with the mayor, so he arranged for the return call to be transferred to that office.

At the library, Dr. Manara went into the stacks to draw out a volume from the massive set of abstracts there. What caught his eye first was that when TCP went over the temperature of 200 degrees C. a substance called TCDD—or dioxin—was accidentally formed. Since the safety valve at the factory had been released in the accident, there seemed little question that the temperature must have exceeded that.

Quickly Dr. Manara began checking and cross-checking many of the abstracts available on dioxin. Fact after fact

26

emerged in the material listed under "tetrachlorodibenzo-p-dioxin," the full name for the dioxin compound. As he read, a chill came over him. The difference between the toxic effects of TCP and those of dioxin is staggering. The abstracts revealed that one gram of TCP in a two-pound mixture would kill half the rabbits who might eat it. It took only 1/100,000,000th of a gram of dioxin to kill the same percentage of rabbits.

One abstract noted: "Dioxin is one of the most potent toxins known to occur as a pesticidal impurity." Another paper read: "It is the most potent small-molecule toxin known," and pointed out that as a poison it dwarfed arsenic and strychnine.

The effects dioxin has upon the human body mentioned in the abstracts were no more encouraging. The liver and kidney can be acutely affected. Lesions of the thymus can develop, which in turn lower the body's immunity. Chloracne, a disfiguring skin disease, can develop, several weeks to four months after exposure. All these conditions are extremely resistant to any treatment and can linger for years, if not for a lifetime. Accidental exposure to dioxin in the past had brought widespread severe illness and fatalities. Animal tests had revealed that slight exposure to dioxin could be teratogenic (could cause birth defects), to say nothing of being mutagenic (able to cause changes in the chromosomes that could lead to cancer). Vietnam studies showed that cases of liver cancer had increased to five times normal after defoliation with even a mild concentration of dioxin in the defoliant.

As Dr. Manara reviewed the medical abstracts, the true horror of the Seveso accident became more and more apparent. The TCP in the poison cloud had presented a disturbing annoyance. The inevitable presence of dioxin created a tragedy. TCP was as mild as milk compared to dioxin.

Dr. Manara later told a friend that the period devoted to

27

reviewing the abstracts that day was "the most awful ten minutes I have ever spent." He added: "I dreaded going back to the phone. Neither I nor the people in Seveso had any idea what an incredibly powerful poison dioxin was. I didn't want to alarm the population, but I had to reveal the truth."

Manara took the material to the phone and called Ambrosi, who was interviewing Mayor Rocca. When Manara's call came in, Ambrosi began taking notes, signaling to the mayor to pick up the other line. Manara began to read portions of the medical abstracts to them. Outside, it was beginning to rain, with several loud thunderclaps. The lights in the office flickered off and on. In the intermittent darkness the reporter and the mayor listened as Manara read off the long record of dioxin and its brutal effects.

Mayor Rocca later said that he felt as if he were in a dream or hallucinating. He could not believe what he was hearing. The rain increased with the thunder. Finally, the lights went out completely. The mayor could think of the day as nothing but apocalyptic, as a scene from Dante's Inferno.

V

WHILE AMBROSI and the mayor were absorbing the shock of the news concerning dioxin, Professor Aldo Cavallaro of the regional laboratories in Milan had already made his way to Switzerland to try to get to the bottom of the mystery surrounding the incident. With him was Dr. Giuseppe Ghetti, the chief local health officer for Seveso. They had been extremely dissatisfied with the reports coming from the Hoffmann–La Roche and Givaudan scientists.

During the ten days following the escape of the poisons, the Swiss scientists had not been idle. But they were facing a critical problem. Both Dr. Adolf Jann, chairman and director of the giant Hoffmann–La Roche firm, and Guy Waldvogel, president of Givaudan, were personally involved in the investigation from the moment the news reached them. They were aware of several past industrial accidents involving TCP—including incidents in England, Germany, Holland, and the United States—in which dioxin had inadvertently been formed. The contamination had been confined to the factories, but the results were harrowing: some lingering deaths among the workers, suspicion of resulting cancer, severe liver and kidney damage, general systemic poisoning, and chloracne. These effects had shown up months, sometimes years after the accidents had taken place.

Beyond all this was the incredible persistence of dioxin. The chemical does not dissolve in water. Once it has penetrated a material, it can remain practically forever. In Holland, the factory where an accident happened had to be dismantled brick by brick, sealed in cement, and buried at sea. In England, a similar factory had been scrubbed and cleaned with infinite care. Eight months later, dioxin had been found in dangerous quantities in the porous walls and floors. Efforts at decontamination had been useless.

At the time of the accident at Seveso, the Swiss officials were faced with determining what the cloud consisted of. In the chemical reactor were a solvent, the TCP (trichlorophenol), and caustic soda. Since the safety valve was not estimated to release until 320 degrees C. was reached, the temperature had moved far beyond the point at which dioxin would be formed. The higher the temperature, the greater the proportion of dioxin would be.

The question presented by the dust and other samples collected in Seveso was: Did the superheating produce *massive* quantities of dioxin, quantities that had never before been dispersed into the atmosphere, that had never before coated the landscape, animals, or householders? There was no precedent.

The detection of dioxin was complicated, requiring slow, expensive processes known as gas chromatography and mass spectrometry. At that time, only the Roche laboratories were equipped to handle these tests.

The first soil and dust samples had arrived in Switzerland shortly after the accident. Within three days, testing of the samples had revealed that there were large quantities of dioxin in the area of the reactor room itself. In an attempt to determine precisely how far and where the pure dioxin might have spread, Roche officials ordered Icmesa to gather samples farther and farther away from the factory. If the dioxin had reached any residential, farm, or business area, there was only one action possible: total evacuation and the complete sealing off of the area—perhaps forever.

As samples had come in during the first week, the picture began to form: the presence of the dioxin did *not* fall off. But the results were still spotty. There were no clearly defined limits of contamination. The dilemma in Switzerland was: Should the local authorities in Seveso and the surrounding areas be informed that total evacuation was necessary? Should an announcement about the presence of dioxin be made before evacuation procedures could be set up?

These questions went unanswered as through this slow and laborious process the Swiss officials attempted to find the limits of the dioxin contamination.

By the time that Professor Cavallaro and Dr. Ghetti had arrived, the Swiss officials had finally decided to notify the Italian authorities that dioxin was present, although the limits of contamination had not yet been determined.

The enormity of the situation was immediately apparent to both Ghetti and Cavallaro. They pressed the Swiss scientists for the full extent of the massive dioxin penetration. The first estimate of the amount released was two to three kilograms, which theoretically was enough to kill more than 100,000 people. Realistically speaking, the poison would have to be administered directly to the people to cause this much havoc, but the threat itself was terrifying.

There was no question now that homes and fields and gardens were contaminated with dioxin as well as TCP. Children had been playing in the grass for ten days. Cavallaro himself had been gathering samples with his bare hands. Most of the local animals and foods eaten over the past ten days were definitely contaminated with the deadly chemical. The situation was critical and catastrophic. Evacuation alone posed a highly emotional and distressing problem. Because of this, the areas evacuated would have to be kept to a minimum.

What suddenly loomed ahead was one of the most complex social and medical labyrinths ever faced by public officials, scientists, doctors, and a community. It was later to be called an Italian Hiroshima.

Once Cavallaro and Ghetti had confirmed that the area was contaminated with dioxin, Mayor Rocca acted quickly. He ordered an emergency center set up in a new elementary school in Seveso, and called on local and regional health officers to staff it. The directors of the Icmesa factory were placed under house arrest. The other towns in the area—Cesano Maderno, Meda, and Desio—were warned that parts of their communities might be contaminated. A state of emergency was declared.

Two major problems dominated the scene. First, this happened to be the prime vacation time in Italy. Milan was practically closed tight. Staff scientists had to be called back from vacations immediately. Medical suppliers had to reopen their facilities. Regional authorities, lab workers, and doctors had to be summoned from their vacations in the Alps, on the Mediterranean shores, in France and Spain.

Second, it was difficult to grasp the deadliness of dioxin without considerable study and reference to the esoteric research abstracts. The subject was extremely technical and scientific. An awareness of the dangers posed by dioxin dawned slowly almost everywhere. This was illustrated dramatically when, in what must have been an attempt to prevent panic, the governor of the region had called a meeting of his health officers on Wednesday, July 21, to assess the situation. TV reporter Bruno Ambrosi, waiting for a news report on the situation, got none.

He finally called the governor directly. Ambrosi was told that there was "no danger" to the situation. Knowing what he did about dioxin, he blew up. "How can you possibly say that?" he blurted out, surprised at his own audacity. He detailed the situation clearly and then asked, "Why haven't you brought the Mario Negri Institute into the picture—right here in your own backyard?"

The call got results. The Mario Negri Institute, with its ultramodern instruments and international staff of scientists, was called into action, along with the extensive regional facilities at the University of Milan.

A giant and grotesque detective story began to form. What massive medical examination program could be set up to test *all* the people of Seveso, plus those in other suspect areas? How far from the factory were the farthest dead animals and contaminated grass? What was the *amount* of dioxin in these? Where was the concentration of dioxin the greatest? Where did the concentration of poison fall off? Where did it disappear altogether? Exactly which houses, offices, and factories would have to be evacuated? Where would people be housed? What experts should be called in? What could be done about decontaminating the areas? What about cars and trucks that stirred up dust and then carried dioxin away with them? What about the adults who had been living in the contaminated areas, eating the vegetables and animals? How much dioxin had they absorbed? What about the children, who had not only eaten contaminated food but played, rolled, frolicked in the grass? What about the dying pets and the birds tumbling from the sky? What about the birds and animals who fled from the area,

33

carrying dioxin with them? Just eleven days after the accident, a nightmare was spreading across several comfortable, prosperous towns, and no one knew when it would end.

An estimated 220,000 people might be affected throughout a wide area. They were not only confused but on the edge of panic. The tension was almost unbearable. Officials, wanting to avoid panic before the lines of demarcation could be drawn and a controlled evacuation planned, were reluctant to talk about the situation. A heavy rain saturating the area brought false hope that the dioxin was washed away. This was impossible. It was such a "hard" chemical that it would persist indefinitely.

The job was literally staggering. All the services of the region's medical, psychological, and sociological units would have to be coordinated to examine, treat, and relocate families under incredible stress. A large part of the job fell on the shoulders of Vittorio Rivolta, the minister of health for the province of Lombardy. An energetic and sensitive man, he was later to describe his activity as a "magnificent obsession." His office was notified about the presence of dioxin in the area on Friday, July 20, ten days after the escape of the poison cloud. From then on, like many of his colleagues, he barely slept. Since it would take considerable time to program the complicated mass spectrometers for detecting the presence of minute quantities of dioxin, the only way to quickly determine who should be evacuated and who should not was an immediate survey to determine where the greatest number of dead animals had been found.

Already the veterinarians in the Seveso area had been collecting them with their bare hands. It was a dirty job, for many were decomposed. One person picked up the body of his cat, and its tail fell off. The viscera of the dead rabbits, chickens,

ducks, and other animals—and this was important—could indirectly provide an index of where the poison was present in
large enough quantities to kill.

When the first carcasses of the animals arrived at the regional institute of veterinary medicine in Milan, Dr. Carlo
Binaghi, director of the institute, prepared to perform the
autopsies. When he made his first incision, he stared in disbelief. The subcutaneous tissues were abnormally swollen. The
fatty tissues of the animal were totally degenerated. The tra-
chea had massively hemorrhaged. The liver and kidneys were
overwhelmed with lesions. "I have never seen such a destruction and devastation of internal organs in my entire career,"
he told a colleague. "It was as if the animal wanted to kill itself
in the most violent way possible."

The devastated rabbit livers suddenly became the most important index in deciding what parts of the land had to be
evacuated—the most urgent priority. People were still living
in an area covered with an invisible blanket of slow but deadly
poison, and no one yet knew exactly where this area began or
ended.

More dead animals were brought to Dr. Binaghi's laboratory
for autopsy; the viscera continued to show the incredible rot-
ting of the liver and kidneys that indicated dioxin poisoning.
The results were immediately passed on to Rivolta at the Ministry of Health office, with precise information as to where the
animals were found.

Meanwhile, Rivolta was faced with the question of the plant
and vegetation contamination. The gross autopsies alone were
not enough to make the critical decision of which families
were to be moved out of their homes, perhaps forever. Some
evidence in animals and plants was showing up to indicate that
the dioxin contamination might extend over an area of up to
sixty square miles.

At the University of Milan's Institute of Pharmacology, Professors Giovanni Galli and Flaminio Cattabeni took on the assignment of analyzing the initial plant and soil samples, in an attempt to outline the critical zones of contamination by these clues. Their tests would be correlated with the first animal findings.

Dr. Cattabeni had heard nothing about the problem until July 19, when he had received a phone call from Professor Cavallaro, who had given him a sketchy background report on the accident and had asked for some help in analyzing foliage taken near the factory.

It took until July 23 for the Milan lab to confirm that dioxin was present in highly dangerous quantities in the preliminary samples. Dioxin was so potent that one part per billion was considered extremely toxic. The first samples showed a concentration of three to four hundred times that. At this point, Dr. Galli had just returned from vacation, and the entire laboratory went on a crash program to meet the emergency. Both he and Cattabeni went to work on a sixteen-hour-day basis, setting up a twenty-four-hour-a-day schedule for more than a hundred technicians. They were to be handling some four hundred tests a week.

Faced with the agonizing decision of ordering the evacuation of an undetermined number of houses, buildings, and factories, Rivolta collected the preliminary data and attempted to correlate the soil and vegetation findings with those of the animals. The results were meticulously posted on a large map in his office.

The importance of the map assembled from the preliminary tests was paramount. The territories of Seveso, Meda, and Cesano Maderno began to be filled up with small red stickers, each indicating where the body of a dead animal had been found. It didn't take long to recognize that a thick cluster of red stickers began to show up in the factory area and extended in a fanlike pattern toward the southeast, the direction in which the poison cloud had moved.

The contamination apparently fell on both sides of the superhighway from Como to Milan, stretching at least a mile southward from the factory. Ominous placards were already posted there for motorists: CONTAMINATED AREA. ROLL UP WINDOWS. CLOSE VENTS. DO NOT STOP. DRIVE SLOW.

The red stickers on the map also clustered around the homes on the Via Carlo Porta and its neighboring streets, over the baseball diamond, recreation center, swimming pool, cemetery, and Seveso's main artery, the Corso Isonzo, which bisected Seveso. The red stickers were tightly packed throughout the area. Rivolta was dismayed to discover that the animal mortality was almost 90 percent in this zone.

Green stickers, indicating sick and dying animals, clustered in two pockets on the east and west of the cone, a mile or so down from the Icmesa factory in the southern part of Seveso. These zones lay along the Cesano Maderno line, where Dr. Ciccardi was mapping his dead and dying animals. Practically all these animals, when examined, also showed the lesions and disintegration of the liver and kidneys that had shocked Dr. Binaghi. Yellow stickers marked locations where animals showed no sickness.

Studying the red, green, and yellow stickers that were spread over the map, Rivolta noted with some relief that the poison cloud had missed the most thickly populated section of Seveso. If the wind had shifted five degrees in either direction, the cone of poisoning would have struck literally hundreds of additional homes and factories. But the present roughly defined zone of intensive contamination was bad enough. It was over a mile long and nearly half a mile wide in Seveso alone.

The area in which the red stickers were packed tightly together was designated Zone A. The area in which the green stickers predominated was called Zone B. And a mixture of yellow, green, and a few red stickers outlined a tentative region to be known as Zone R—the Zone of Respect. In consultation with regional, national, and local authorities, Rivolta reluctantly ordered the total evacuation of Zone A on Saturday, July 24.

Over forty families and more than two hundred people would have to be ordered out, two weeks after the disaster had struck. Among them were the Zorzis, the Cassios, and all the families along the Via Carlo Porta. Seveso's police chief Calo, summoning up his courage, joined with the carabinieri to handle the dreaded job of notifying the families that they all would have to leave—each with only a suitcase and the clothes on their backs. To Calo, it was one of the most painful assignments of his long career.

VI

THE PROBLEM with dioxin was that it was noiseless, odorless, tasteless, and invisible. The authorities and the people were faced with an enemy they couldn't see and had no means of fighting. The enemy occupation was an accomplished fact. The houses along the Via Carlo Porta, built with such love and care, looked no different than they had before. No sign of the white crystals or the waxy, oily coating remained on the greenery. One housewife whose four-year-old daughter's body was covered with running sores and pustules said, "I think I'm going mad. We fear everything. Maybe the children will come down with something else; maybe they will die. If only we could *see* dioxin. It's so much worse because it's invisible."

The order to evacuate Zone A came so quickly that few were ready for it. The Zorzis and the Cassios, along with their neighbors, watched the police and carabinieri approach, the latter in battle dress but walking with unheroic sadness. They gently, even tenderly, guided the people into the cars and buses. No furniture or utensils could be taken—just the clothes on their backs and a suitcase.

There was no resistance. A child cried as she left her two kittens behind. Mrs. Cassio's eyes filled with tears as she looked

39

back at their house, wondering when or if she would ever see it again. Although released from the hospital, she was still in pain; her stomach was still swollen. The Cassio car joined the others moving in a slow cortege toward the center of town, where the families would be assigned to a hotel near Milan for an indeterminate time.

The two hotels were chosen with care by the regional authorities, with an eye to making the abrupt adjustment as painless as possible. Both were modern, with comfortable accommodations and surroundings. The Motel Agipe was a standard modern facility, part of a chain that catered to tourists and businessmen on the road. The Leonardo da Vinci complex was large, impressive, and quite luxurious, with a large swimming pool and handsome lobbies. If it were not for the agonizing tension and uncertainty hanging over these people, and their being wrenched from their homes with little or no warning, the hotels could have served as a pleasant vacation respite. Instead, they quickly became velvet prisons full of fearful people concerned about being sentenced to excruciating disablement or even death by the poison cloud they had lived under so many days. The prognosis so far was limited to wild rumors and deep foreboding.

Those who were not evacuated were no less anxious than those who were. Rumor had it that the danger zones would be far extended; no one felt secure. The thirty patients in the hospital, most of them children, were not responding to treatment, which even seemed to make the symptoms worse. Word that there was no known antidote for dioxin poisoning was quickly passed around, adding to the psychosis of fear hanging over the area.

The confusion was punctuated by the difficulty in grasping the indestructability of the poison. One elderly man reassured a local health officer that he had boiled all tomatoes he picked from his garden two times, not realizing that boiling would do nothing to destroy the dioxin. Another man ate ten of his twenty-six rabbits that had died, still thinking the warning applied only to fruits and vegetables. A woman cried when she was told she would have to destroy her strawberry jam.

Shortly after the first evacuation, an army battalion arrived and began rolling out barbed wire—six miles of it—to fence off Zone A. Sentries were posted. No one was allowed to enter unless dressed in a special suit and mask. Included were the veterinarians and volunteers moving in now on an assignment to kill every living animal in the newly designated Zone A. Roadblocks were put up to mark off Zone B, where the animals were also ordered to be killed.

It was not a pretty picture. Moving past the armed sentries, the men entered a ghost town, catching chickens inside the fenced pens, seizing the rabbits in their hutches, and twisting their necks swiftly. Then they dropped the animals into black plastic garbage bags to be shipped to the laboratories in Milan for testing. Gradually some 50,000 animals were killed this way, to avoid the spreading of the contamination and to provide an index as to what the limits of the final evacuation zones should be. Ducks, sheep, goats, and other livestock similarly had to be tracked down and slaughtered; eggs, cheese, and milk were destroyed. Hunters were warned not to eat their game but to turn it in to the health authorities. Beehives and honey were destroyed up to three miles away from the contaminated zones.

Moving into Zone B, where families were still permitted to live, the reluctant executioners met with no opposition, only sadness and despair. These people depended on the animals and crops they raised for most of their food. Yet many mutely assisted in the slaughter, frequently offering the veterinarians and volunteers a modicum of hospitality and a drink. The unspoken question in their minds was: Will we be the next to be forced out of our homes?

The question was a valid one. Rivolta's map was showing more and more animal fatalities clustered in Zone B, as well as dioxin concentrations in the plant and soil specimens. The prospect of ordering more people out of their homes was abhorrent to Rivolta. But by July 29, when 1,400 soil examinations were in, it became obvious that he must enlarge Zone A. This time nearly six hundred people were forced to leave their homes. The psychological consequences of the enlargement of Zone A were harrowing. In a series of anguished phone calls, people were demanding to know why they couldn't take more of their belongings with them, how long they would have to stay away, what the ultimate fate of their children was going to be, and if it was certain that they were going to die. Beyond that, there was the frightening question of damage to the fetuses of pregnant women.

For the mothers in Zone B, the first phone calls from the social workers came as a shock. All their children, without exception, were to be taken away to the relative safety of resort camps, some for boarding, some for day care. None could be excluded. The news was horrendous; the mothers were stunned. Many refused to cooperate.

"It is unacceptable," said one distraught mother. "It is as if we are in a war. I will fight you. You cannot take my child away."

"I am sorry," said the social worker on the other end of the line. "We have to come and get them, regardless of what you say."

And they came. Some families locked up their children and barred the doors. Others stood mutely or cried. Some even screamed at the workers. But eventually they gave in. The children were wrenched from their parents. The sounds of the birds, the animals, the children had now all vanished from Zone B.

Pushing hard for more information, Rivolta followed closely the operation at the Mario Negri Institute, whose instruments were capable of determining the dioxin content more precisely than instruments at other labs. Recognizing that the amount of dioxin released had the capability of killing more than a billion guinea pigs, the entire scientific staff at the institute voluntarily canceled vacations to cooperate with the government. Under the direction of Professor Silvio Garattini, the highly trained staff rearranged much of their entire research program to set up for the delicate job of pinpointing the amounts of dioxin in the animal livers sent to them by the government teams at the University of Milan. The shiftover was a job that would have taken months to do under ordinary circumstances; it was accomplished in a matter of days.

The techniques they used were sophisticated and subtle, but their findings would have a critical impact on the lives of the people who were suffering from agonized uncertainty. Special precautions had to be taken. An entire special building was constructed in one day to handle the deadly solid and liquid garbage from the dioxin-infested materials. A completely new laboratory was constructed in a week. Only one obscure scientific paper had been written on how to prepare a tissue extraction for tracing dioxin; the process took two days and two nights to complete. Mario Negri scientists found a way to cut the time to a fraction of that.

Working with such a precarious poison allowed no margin for error. Two grams of an animal's liver had to be dissolved in a phosphate buffer solution, and then spun to 9,000 G's in a centrifuge. Then a supernatant fraction—floating on the surface of the solution—had to be centrifuged again at 100,000

G's to produce a solid pellet left in the bottom of a tube. Portions of this in turn were introduced into the mass spectrometer, an imposing $300,000 instrument with sophisticated electronic circuitry that could detect if dioxin was present and exactly how much (down to 1/10,000,000,000,000th of a gram).

There were only five mass spectrometers in all of Italy. Three of these were in Milan. But so urgent was the need for an additional mass spectrometer in the face of the crisis, that a special instrument was ordered from Sweden. It was packed on a truck at nine in the morning, ferried across the Baltic, and driven down through Europe and through the Simplon Pass into Italy to arrive by seven o'clock the next night, a process that would normally take six days.

In disposable paper uniforms, masks, and gloves that were resistant to the solvents involved, the Mario Negri scientists had to trace the poison indirectly by looking for certain enzymes known as glucuronic acid transferase and diaphorase. It is known that dioxin increases the quantity of these enzymes, which in turn creates the presence of a substance called P-448 cytochrome, the "fingerprint" of the poison. In this way, a code for determining the zones to be cleared could be defined by the tests.

Each small specimen had to be analyzed on the molecular level. Diluted in alcohol, the specimen was sucked up in a syringe and squirted into the injection port of the mass spectrometer. Here it was evaporated by a flash heater and bombarded with tiny bullets, as helium atoms of the carrier gas pushed the molecules through the instrument. Some molecules moved faster than others, and those of like substance were separated. There was more bombardment in a high vacuum area, this time by high-speed electrons. Moving at speeds of up to 5,000 miles a second, the different types of molecules separated from one another, to be recorded on chart paper identifying each type of molecule. Later in the process, the exact amount of each of the substances was recorded on photo-sensitive chart paper to complete the analysis.

VII

THE DATA coming into the health ministry's office from all sources was being analyzed almost hourly by Rivolta, and the picture was not good. The map of the area, with its growing number of varicolored stickers, looked like a bad case of measles. By July 29, Rivolta was forced to enlarge Zone A again and order another evacuation where the tests indicated the presence of as little as five micrograms of dioxin in a square meter of soil. With the order, a total of nearly 1,000 more people were forced to leave their homes, while another 5,000 in Zone B wondered whether they would be next. Beyond that, another 8,000 in the neighboring town of Desio waited in suspense as the results of soil samples being taken there were analyzed.

Meanwhile, more barbed wire was unrolled by army units to enlarge the zone, and the ghost town grew. The sight chilled even seasoned reporters who had recently covered the devastation caused by a recent earthquake in northeastern Italy. Aside from an occasional dead cat or bird, there was almost no visible damage. The houses standing behind the barbed wire were unscathed, sitting in silence with their blinds rolled down. A huge construction crane sat beside the skeleton of a large apartment house under construction. The grass was still lush and green and the trees were bursting with fruit. "It looks like a scene from a weird science fiction story," said one reporter.

Not the least pressing of Rivolta's duties was trying to discover what could possibly be done to decontaminate the land and houses of the entire region. The cloud had skipped over some areas in an updraft, and had come down hard on others farther out. Animal and soil tests clearly showed this erratic pattern. Zone A and Zone B sprawled at least two miles downwind through Seveso and Cesano Maderno, then leapfrogged over to Desio.

Decontamination efforts in past factory accidents had been futile. Two pipe fitters who had returned to the factory three years after the British accident came down with severe chloracne after they worked on a reactor that had been thoroughly steam-cleaned. Even though they wore masks and special suits, Dutch workers attempting to spray-clean a contaminated area contracted the disease, and some passed it along to their families. One wife recalled: "It was dreadful. I had never seen anything like it. The men had big, black holes in their faces, which gave out an unbearable stench."

None of the decontamination specialists consulted by Rivolta could offer much immediate hope. Alex Rice, an expert from the British firm of Cremer and Warner, felt that the greatest concern was to prevent the dioxin from spreading. This was a highly likely possibility unless drastic steps were taken to prevent it.

He arrived in Italy at a critical moment. A suggestion had been made to bring in a chemical-warfare unit from the army with flamethrowers, to burn off the grass and foliage. This would have brought automatic disaster. Far from being destroyed at this temperature, the dioxin would have been

47

spread across the countryside by the smoke and flames. Dioxin can only be destroyed at temperatures of over 1000 degrees C., requiring a specially built incinerator.

Dr. Anne Walker, a dermatologist who had treated workers after the British accident, joined other scientists in the opinion that the long-term effects of dioxin exposure might not show up for ten or fifteen years. In a brutally frank statement to the press, she said: "I would evacuate everybody in the contaminated areas at Seveso, and not allow them to take anything with them—not even their clothes. Then I would seal off the area, and leave it barren, if necessary forever. Only in this way could one be certain of preventing further contamination."

The greater the number of suggestions that reached Rivolta's desk, the more desperate the situation seemed. The recommendations of a U.S. Department of Agriculture specialist pointed up the enormity of the job. They were as follows: (1) force Hoffmann–La Roche to buy all land in Zone A, including buildings and animals; (2) require a nine-foot-high wire-mesh fence, lined with plastic or canvas; (3) require a high-temperature incinerator, and construction of a sixty-foot-deep disposal pit to bury all buildings, plus all asphalt from roads in Zone A and B; defoliate all trees, along with 1000 degree C. incineration of leaves, and so forth.

The recommendations as to what *not* to do were equally important. In addition to the dangers of low-temperature burning, the scraping of the topsoil was equally risky. There was nowhere to put it.

The consensus of most experts was that the area should be sprayed with an olive-oil mixture, which if exposed to the sun might degrade some of the dioxin. Givaudan was preparing specialists and equipment to do this. But several experts felt the entire contaminated area should be sealed off forever. The porous surfaces of the buildings would probably yield to nothing; dioxin in their hidden cracks and crevices would never be exposed to the requisite sunlight; no one living in the areas could ever feel secure. Rivolta, his desk piled high with requests that called for urgent attention, forlornly summed up the decontamination question with: "Nobody has any clear idea how to decontaminate the area, or how long it would take."

VIII

THE FACT that there is no known antidote to dioxin poisoning made Rivolta's job of organizing a medical program for the thousands of people who had been exposed to the chemical full of despair. The immediate first-aid steps to alleviate the early symptoms of TCP were relatively simple, but the long-term effects of dioxin remained unknown. The fear and uncertainty that spread among the people were as great a problem as the gloomy prognosis.

Some five hundred patients had been afflicted with such problems as painful skin lesions and liver and kidney malfunctions. Carefully monitoring the people who might have been directly exposed created a staggering medical problem, involving a battery of more than twenty exacting tests for each person, requiring the most sophisticated diagnostic techniques. Thousands of people jammed the emergency center at the new Seveso elementary school to give blood samples that would have to be analyzed. Dr. Ghetti, who barely had time to sleep since the crisis began, worked tirelessly to try to meet the emergency, but along with everyone else he was working in the dark. It was impossible to answer all the questions fired at him by the strained and anxious townspeople. Almost in despair, Dr. Ghetti told the press: "We do not know exactly what we are up against. No one can state with certainty that no deaths will occur."

As hundreds filed by the school tables during the first few days, the small test tubes filled with blood formed a mountainous pile, with each specimen carefully labeled and identified. If it were not for a fortuitous coincidence, it is unlikely that such a flood of material could have been handled. As it was, with the aid of nearby Desio Hospital, Dr. John Tognoni of the Mario Negri Institute had recently set up a computer program for specialized pharmaceutical analysis. The hospital already had modern and sophisticated laboratory equipment, and combined with the recently installed computers, the hospital's capacity to handle and analyze the blood samples was increased measurably.

The battery of tests for each patient included an analysis of the liver function, kidney function, bone marrow, hemoglobin, albumin, and other key tests to trace the effect of dioxin on the system. Most important was to determine if the number of lymphocytes in the blood was reduced—a sign that the body's natural defenses were lowered.

At the modern, impressive Desio Hospital, Dr. Paulo Mocarelli, recalled from vacation by Rivolta on July 25, went into immediate action to handle the crash program. On July 26, 267 persons were examined for a total of 6,000 tests. By July 28, 1,577 patients had been examined, for a total of nearly 30,000 individual tests. The crash medical program's ultimate aim was a long-term focus on 8,000 of those who had received the most intense exposure, who would be known as "risk people." Their tests were to be repeated every year, with special attention paid to impaired immunology, as signaled by a low lymphocyte count. Lesser controls would be applied to the population of 200,000 dispersed throughout the outlying areas.

Working late into the night, Mocarelli and a staff of twenty-seven technicians assembled and checked the long computer sheets to correlate the results from Zones A and B, and the peripheral "Zone of Respect." Mocarelli found the staff working with a sense of dedication he had never observed before, in spite of sheer physical exhaustion from lack of sleep. Special attention was given to children of Zone A, who had played many days on the grass, the fields, and in the community swimming pool, where they often swam from morning till night. Their reaction to dioxin might take one to two years to show up.

IX

TIED INTO the medical program was the importance of making people aware of the precautions that they had to continue to take. All residents of Zone B, under the shadow of future evacuation if more precise tests so indicated, had to be continually reminded not to touch foliage or the ground, not to eat anything they had grown or raised, and to send the children out of the zone to play in specially designated areas after their return from camp. Their homes became de facto prisons much of the time. The toxicity of Zone B became rather glaringly apparent when a young man went out to a field at night with his girl to make love. Within a few days, his back and arms became covered with lesions.

One of the most ominous threats hovered over the women who were pregnant. They faced the agonizing possibility that their babies might be born with deformities, as in the thalidomide crisis of years before. But these women experienced a deep inner conflict of whether to have their babies or not, since the Italian government and Catholic Church have historically been opposed to abortion and contraception. In Italy, abortion for any reason was a punishable crime. Shortly after the Icmesa accident, Italy's new Christian Democrat minister of health took steps to facilitate abortions in the contaminated area, based on a 1975 court decision that the old law was partly unconstitutional. An abortion could be performed if the physical or psychological health of the mother was in danger.

The Church, however, remained firm in its conviction that abortions should not be carried out. It maintained that the purpose of the abortions would be to destroy the fetus rather than to save the life of the mother. The archbishop of Milan asked for volunteers to come forward to adopt any deformed children born of parents who were unable to take care of them. Monsignor Giovanni Guzzetti, director of family affairs for the archbishop, softened the stance somewhat by saying that the Church would not pass judgment on women who decided to get an abortion, although the Church emphasized the importance of the life of the child in any condition. The Church, Guzzetti said, was asking people to stay close to those who did have abortions because they would need love and help. Meanwhile, the Church would continue to remind wives and mothers that it was their moral duty to use natural rather than artificial means of contraception.

The prevention of conception was on almost everyone's mind. Doctors were warning couples not to conceive, at least not until the medical picture was clarified. Semen was being examined for the possible effects of dioxin. More than a hundred expectant mothers sought advice from the health-care centers in the area. The doctors presented them with the facts: laboratory animals display embryo malformations at dioxin concentrations as low as .05 parts per million; the effect on the human fetus was a question mark. But the decision was left up to the pregnant woman, who was torn between her own conscience, the Church's dogma, and the ambiguous laws of the country.

It was not an easy decision to make. One expectant mother

came out of the health-care clinic and said grimly, "Isn't it just as sinful to bear a malformed baby that will spend its life in an institution as it is to have an abortion? It's much more humane to abort."

Another was bitter. "The birds are no longer in the sky," she said, "and there is no cure whatever for monsters."

X

ADDING TO the harrowing woes of the area was the crushing economic impact. Seveso was in the center of the Brianza region, known for its quality furniture-making. The artisans and craftsmen took inordinate pride in their work. Not long after the event, word came that huge shipments were being held up at the Swiss and German borders, whether the furniture came from the contaminated zones or not. Enormous inventories of select lumber lay exposed to the sky; the presence or absence of dioxin made no difference. Everything was suspect if it came from anywhere near Seveso.

By mid-August, a special scientific commission set up in Rome despaired of any reclamation efforts in the contaminated areas. Dr. Aldo Cimmino, who headed the group, called for total defoliation and added: "It is useless to give any hope to the residents of the area. The buildings are all condemned. For now, we should not destroy them because of the problem of dust. It's better to wait three years, when the concentration of dioxin will be reduced." At the same time, the Italian cabinet under the new government of Andreotti's Christian Democrat regime allocated some $48 million to meet the rapidly mounting cost of emergency health care and resettlement.

In Basel, Switzerland, Adolf Jann, chairman of the board of Hoffmann–La Roche, stated that the company would pay for all material damage resulting from the accident, and would work with the Italian government to try to neutralize the effect of the poison.* Along with Guy Waldvogel, chief executive of the Givaudan subsidiary, Jann countered the charge that the company was slow in letting people know about the presence of dioxin by saying that at the time no one knew exactly what the poison cloud contained. "There was no measurement of any zoning to know what to do about it," Waldvogel said. "In the middle of the first week, we mentioned that highly toxic trace products could have escaped. Dioxin wasn't mentioned at first, because we didn't know it was there. It wasn't until July twenty-third that we had enough measurements to draw a map, which we handed over to the Italian authorities."

But should the safety valve have been designed to vent directly into the atmosphere without a dump tank or containment vessel? Waldvogel had little to say. "We cannot at the present time really comment on that," he said, "because the whole thing is under legal proceeding. The only thing we know is that we don't yet have a real explanation for what happened, but we have reason to believe that a combination

---

*Last fall the Hoffmann–La Roche/Givaudan group created a fund of $11,500,000 to be distributed as compensation for damages to individuals and firms in the contaminated area as well as to Italian regional authorities. An office was created in Milan to handle compensation payments. But most consider the damages to be inestimable.

of technical and human error was involved."

Waldvogel saw some relative encouragement in that most of the severe early symptoms were confined to the residents of the houses nearest the factory. "Those houses," he said, "were built by the workmen themselves—not by Icmesa workers. By all industrial standards the piping systems and general personal hygiene of that group of people is relatively bad. The way the people live there, about washing, bathtubs, and things like that, are not comparable to what you might expect."

Apparently Waldvogel had never had a chance to visit the homes of the Zorzis or the Cassios—or those of any of their neighbors. Mrs. Cassio, for instance, took immeasurable pride in her house and furnishings; she even had little wool footpads at the entrance to each room.

Meanwhile, the displaced families were living in a numbed and confused condition in the hotels they were assigned to, in relative physical comfort but still stunned from the abrupt disruption of their lives. They found it hard to adjust. They missed all their personal things. Always, there was the thought of what was going to happen to their health, and that of their children.

One still-unanswered question was how any company, of whatever size, could possibly pay for their loss. The material destruction was staggering enough. But the direct and indirect economic losses were almost impossible to estimate. It was doubtful that any insurance policy issued by any company could handle such losses, and Hoffmann–La Roche was not disclosing its insurance coverage. The company was paying for the hotel costs of the displaced in full, and it had offered to compensate the farmers for their lands and their crops, and to continue to pay its employees their wages—but this was at best a Band-Aid treatment.

Britain's prestigious journal *The New Scientist* maintained that the accident had world-wide implications. In two articles on the Seveso tragedy, the publication questioned the balance between the benefits and risks of such chemical enterprises. "Not only is the dioxin released over a huge area of Milan's suburbs one of the most frighteningly toxic substances ever made, but the whole purpose of the Seveso plant (and others like it) has been thrown into the melting pot as the hazards of its end products are reassessed," one article stated. It added that the release into the atmosphere of an estimated four and a half pounds of dioxin was catastrophic.

The article brought up further questions that were continuing to burn in many minds: "Who needs trichlorophenol (TCP)?" it continued. "Both hexachlorophene (for surgical soaps) or 2,4,5-T (the defoliant) should be replaced with safer substances anyway."

This was a rational question. TCP's offspring, hexachlorophene, had killed thirty-seven babies in France, when their bodies had been smeared with a talc base that contained an accidental 6 percent hexachlorophene rather than the usual 0.1 to 0.2 percent. Other babies had gone into convulsions from pHisoHex, a "normal" solution. Regular use of deodorant soaps containing a mere 1 percent hexachlorophene could build up a cumulative residue of the substance in the bloodstream.

Beyond this, *The New Scientist* asked the question that was growing more pressing every day; "One question that Hoffmann–La Roche has to answer is: why did it run a plant that had a safety valve venting to the atmosphere when it must have known, at least since 1971, that in an accident one of the most potent teratogens ever discovered could be spewed over the surrounding population?"

Like Waldvogel, Roche's president Adolf Jann was not pre-

59

pared to answer the question before the legal process began. He did, however, say, "Accidents like Seveso can always happen unless chemical plants are closed, and that goes too far."

Meanwhile, Donald Lee, a British chemical pollution expert, was continuing a survey for the Italian authorities, examining the data and trying to come up with some clear recommendations. Without any official announcement, it was reported that he was convinced up to 260 pounds of dioxin must have been released. This was up to fifty times the original estimate, and enough poison to kill 12 billion guinea pigs. What it might do to people remained the tormenting question.

XI

IN SEPTEMBER, two months after the accident, panic among the people had turned to frustration and slow, surly anger. Although ninety-four out of every hundred rabbits found dead in the contaminated zones had received lethal doses of dioxin, no humans had died yet, and so far the people had only vague and nonlocalized symptoms to report. Internal pains were frequent, and some began noticing blood in their urine; serious skin sores persisted. The telltale sign of dioxin poisoning, chloracne, had not yet shown up. The initial sores were attributed to burns from TCP and the caustic soda in the mixture. Chloracne has a long incubation period—up to three or four months. Since chloracne is such a clear-cut indication of dioxin poisoning, its occurrence could be a sign of internal damage, as well as being a condition that will linger over years.

The results of the blood tests accumulated since the disaster were not encouraging. Out of 10,000 people tested, nearly 1,000 revealed a lowered lymphocyte count, indicating fewer white blood cells and reduced ability to deal with infection. Even more seriously, the low count could signal damage to the bone marrow, which in turn could eventually lead to leukemia. All the children in Zone A had shown this decrease.

Fifteen women had applied for abortions; three had gone down to the Mangiagalli Obstetric Clinic in Milan to undergo the procedure. Psychiatrists had certified that bearing the babies under the dioxin threat would cause them lasting psychological damage. The Church was distressed, but one expectant mother said: "It's all right for the priest. He doesn't have to have any children. I'm a Catholic, but I'm going to have an abortion."

Even the doctors faced a painful inner conflict. "I am aware that in doing this, we are interrupting a life," said Dr. Giovanni Candiani, director of the Milan clinic. "I am against abortion, and I will never agree to free abortion. But I decided to take the responsibility, even though it upsets me morally."

Gynecologist Francesco d'Ambrosio of the same clinic was less torn, more bitter: "The Roche Company knew from samples taken back to Switzerland that dioxin poison was present within forty-eight hours. And they didn't come here until at least ten days after the accident to report that it was dioxin. It is a known fact that deformation of newborn babies results from exposure to this chemical. We have precise examples in Vietnam."

There was a question of who was worse off, the Zone A families confined in their hotels, or those in Zone B, who lived at home in semi-imprisonment. The streets of Zone B were silent. The children were returning from camp, but still sent away to play during the day, and the families missed the sound of their laughter. All shoes had to be cleaned before entering the front door, but how could one be sure they were clean, and how could the rags or paper towels be safely disposed of?

One child from Zone A returned from a resort camp to tell his mother: "Mama, I watched the soldiers and the barbed wire on TV. I saw our house. It was empty. Is it still there?"

His mother assured him that it was. "Please take me to see it. Please. I want to be *sure* it's there."

The mother acceded. She drove to the carabinieri outpost, where a sentry stood. She begged for permission to simply drive by her house with her boy. It was deep inside the zone. The sentry refused. No one could enter, especially with a car. Perhaps if she walked? He might look the other way.

But the house was too far to walk with the child. Then she saw a cyclist coming toward them. She begged him to loan her the bike just for the few moments it would take to ride by the house. The cyclist did so, waiting by her car. The mother hoisted her child onto the handlebars. The soldier looked the other way. Within minutes she was back. She returned the bike, nodded gratefully to the sentry and cyclist, and left.

The people continued to bring their dead animals in paper sacks to the basketball court of the local school. They held them like bags of groceries, dumping them grimly on the table and retreating silently. Veterinarians carefully tagged and recorded the carcasses and threw them unceremoniously into a large freezer for shipment to the Milan laboratories. There was a sense of quiet despair in the hollow gymnasium, a sense of resignation.

In another corner, white-suited technicians from Givaudan sat sipping coffee on a break from their decontamination chores. All day technicians tracked in the dust and dirt on their enormous heavy boots, apparently unconcerned about the hazardous nature of their tasks. When someone pointed out to one man that he had a wide gash in his protective uniform, he shrugged and poured himself another cup of coffee.

Maps on portable bulletin boards in the center of the room were pockmarked with red stickers, and Zone B had grown from some one hundred acres to almost seven hundred. But no more people were evacuated. Health minister Rivolta had decided that the psychological distress of further evacuation would be worse than the physical risk of remaining in the Zone B homes.

Even the dead were affected. One September day a funeral procession moved to the gates of the cemetery, where the mourners were stopped by the carabinieri. Only the pallbearers were permitted to carry the coffin to the grave, which was just over the line in Zone A.

To add to the distress, a cow which had grazed only briefly in the Seveso fields died from leukemia; she had previously given birth to a stillborn calf. Half a dozen other calves from the same barn had also been born prematurely. Local health director Ghetti commented: "This is one more piece of evidence that will brand Seveso as another Hiroshima."

And as if to punctuate this, Pasquale Mollica, a forty-nine-year-old worker at the Icmesa factory, was taken to the Milan hospital and operated on for cancer of the liver. He died several weeks later.

XII

AT THE Hotel Leonardo da Vinci, a young psychologist sat at a table quietly interviewing some of the displaced persons from Zone A. Children ran and played among the sofas and the chairs, paying no attention to the clusters of black and scarlet pockmarks on their faces. But in spite of this disfiguration, no clear-cut diagnosis of chloracne had been made by early fall.

The Cassios and the Zorzis, memories of their happier days on the Via Carlo Porta still lingering in their minds, chafed at the forced inaction at the hotel. The fear of what would happen to their health was with them always. A few days before, Mrs. Cassio had overheard a doctor say: "It won't be too long before they all will be falling like grapes."

"We try not to speak about this," Mrs. Cassio said to a visitor. "The doctors have already told us that we have accumulated this dioxin inside us, and that it is a dangerous situation. It was a gradual death for the animals. We don't know how eaten up inside we are. We know we will never get back in our home again. We haven't received any money yet for rebuilding it somewhere else. We made sacrifices, and we want houses. We were all hard workers with lovely little homes. We were quiet.

We did our work. We didn't bother anyone. Anyway, we are afraid to go back even if they did let us. If we wanted to sell our homes, no one would buy them. Zone A is a desert, left only to the rats."

Gino Zorzi joined in the conversation. "What has happened to our respiratory systems?" he asked. "We have seen the animals die in front of our eyes. We have breathed the same air they did—for two weeks before we were moved out. Why won't they tell us the effect of dioxin on humans? Are they going to tell us whether the dioxin has destroyed us internally or not? We just learned of a doctor who said, 'These people have already been condemned.' "

When the visitor suddenly burst into tears, Mrs. Cassio asked, "Why do you cry? Do you know some information we don't know?"

"We have to laugh at ourselves," Mr. Zorzi broke in. "Try to laugh with us. We don't want to dramatize anything. We just want to go on with our lives.

The yearning to see their own homes again built up to an obsession. Under pressure, the authorities had given special permission to the Cassios and several other families to retrieve certain specified objects that had been kept in closets and cabinets, where it was assumed that the dioxin had not penetrated.

The Cassios went with the others to the guardhouse by the barbed-wire border of Zone A. They put on white suits, like space men, along with masks, gloves, and boots. Under military escort, they were led to their houses. As the Cassios entered their front hall, Mrs. Cassio burst into tears. She asked her husband to hurry; it was too much to bear. As the soldiers watched, Mr. Cassio went to the closet and took down the heavy bags containing winter clothes. The soldiers nodded approval. Mrs. Cassio went to get Carlo's textbooks. They had been closed in cabinets and were also approved. They looked around their house, at the spotless parquet floors, the television set, the living-room furniture, the immaculate kitchen, the tiled bath. Everything was just as they had left it. They left immediately. At the guardhouse by the barbed wire, they went to special tents and removed their uniforms to be burned. Their possessions were wiped down by the police, and the car was taken to be washed.

At the Leonardo da Vinci, young Carlo Cassio had become spokesman for the evacuated families. Intense and energetic, he was consulting with lawyers to press for immediate repayment for their lost homes. In mid-September the people from Zone A met in Milan with Assessore Peruzzotti, the Lombardy minister of housing. Everyone's nerves were on edge; no one really seemed to trust anyone else.

Peruzzotti told the Zone A residents that members of his department had been out hunting for apartments; rent and expenses would be paid by the government for an indefinite period until permanent housing could be found. Television sets, refrigerators and washing machines would be supplied; a furniture allowance of 70 percent of the declared value of old furniture would be provided. He hoped to get all the people into apartments by the end of September.

But the people bitterly protested that 46 of the 175 apartments that had been found thus far were near a factory producing chemicals that created a constant stench in the air. One man in the audience rose and demanded some definite assurance that he could get back into his home. The minister of housing said he could give no such assurance, and that it might be banned forever. Another man asked if it was true that the people would die in six months. The minister, who had been under constant pressure for weeks, could not answer. He turned and left the room.

It was obvious that no one could answer the question. It was also obvious that nerves were so raw that logic was lost. The psychological damage was as heavy as the medical. By the end of September, the frustration of the displaced people was so great that their emotions were building to a breaking point.

XIII

THE LEONARDO DA VINCI and Motel Agipe grew into bitter pressure cookers. In the cafeterias, the Zone A families clustered with solemn faces and lowered voices. The talk was always the same. The decontamination process appeared to be stymied and ineffectual. Dioxin, they knew, had entered every cranny, every porous surface, every crack, every leaf, bark, all the grain, hay, grass, and soil. Basically, there was doubt that it would ever disintegrate. The only purpose of decontamination now was to prevent the poison from spreading out of the condemned desert of Zone A—and probably parts of Zone B and the Zone of Respect beyond that.

No one knew with total accuracy where the dioxin had spread or penetrated. Carrion eaters had devoured carcasses containing dioxin and flown away. They had no respect for artificial zones. Dust had been kicked up by trucks and military vehicles. Violent winds sweeping down from the Alps had whisked soil and particles to unknown areas, perhaps miles away. Birds, wild animals, and rats had left the contaminated zones for parts unknown.

The special multimillion-dollar incinerator designed to burn material at 1200 degrees C. could not be completed for three or four years. Even if it were available immediately, the community was now in fear of it spreading more poison. Many dead carcasses in plastic bags were piled behind the walled cemetery, with no means of disposal yet available. Another proposal to remove the topsoil was contested by experts, and the protection of men who entered with bulldozers was uncertain. A proposal to use special microorganisms to degrade the dioxin showed only spotty success in experiments.

Autumn leaves, saturated with poison, would have to be sprayed with a chemical glue to keep them from spreading, and of course they could not be burned by conventional incineration. Parts of a nine-foot fence were constructed, but its effect on the residents was marked and depressing. Heavy rains had pushed the dioxin down further, from a few inches to nearly two feet. Traces of dioxin were found in riverbed sediments.

The first symptoms of classic chloracne from dioxin had at last begun to appear: masses of straw-colored cysts, pustules, abscesses, large islands of blackheads known as comedones were added to the lesions earlier caused by the TCP poisoning. The cases mounted to nearly forty; the treatment was savage. Hot towels were placed on the skin, followed by scrubbing with coarse cloths until the pus was removed and blood appeared. Then cortisone cream was applied, and the face was covered with a full wrap-around mask in which holes had been cut for the mouth and eyes. The treatment itself was so painful that some of the children had to be anesthetized before it. A few skin grafts were performed.

Three months had gone by since the cloud had drifted silently over the towns and countryside. At the Motel Agipe, more tightly packed than the da Vinci, passions were running high among the five hundred and more evacuees. The rooms were overcrowded. Tensions were increasing. The promised new quarters and funds for damages were nowhere in sight. Meetings were held almost daily in the motel's conference room. Lists of grievances were made and presented to officials, but went unanswered. There were signs that the bubble would burst, that reason and caution would be thrown to the wind.

At a meeting on October 9, the mood of the group of despairing residents became surly, almost unmanageable. They wanted action, and they wanted it fast. No one knows quite who suggested it, but the idea seemed to grow spontaneously, rising out of the rumblings at the meeting. There were shouts and demands, but those had been common, and no officials were there to respond to them. They were beginning to disbelieve the authorities, anyway. It seemed impossible that the poison could still remain in the contaminated Zone A. They were convinced they were being misled, that the danger was exaggerated, that their homes were surely safe by now. And what about the fateful ten-day delay while their children romped in the poisoned fields? What could be worse than that? And didn't the autostrada cut straight through Zone A from north to south? And weren't vehicles moving on it by the hundreds, every day? What about the Corso Isonzo, which cut through Zone A from east to west? Why was that barricaded, when the superhighway was not?

The decision of the meeting was to act—and to act in the most dramatic possible way. There was no need for a vote; the plan was adopted by wild, almost hysterical acclamation. But with the shouting, there was resolve. The plan must be kept secret. No one must know about it until it happened—the next day, October 10, the day that marked the end of the third month since the escape of the poison cloud.

All but the elderly gathered early that morning outside the

71

Motel Agipe. They talked in low, soft tones now, the hysteria of the night before gone, replaced by bitter resolve. Outside the motel were the two regular buses provided by the regional government to take the residents for daily visits to Seveso. Their routine schedule called for a seven-thirty departure. The drivers sat in the buses, waiting quietly.

There was a larger crowd than usual on this morning. Within moments, both buses were packed. But beyond that, an inordinate number of people were crowding into the cars in the parking lot. The first bus contained a much larger percentage of children than usual. There were few greetings among the people, and the children were exceptionally quiet. At precisely seven-thirty, both buses started out of the motel driveway. They had not gone far when an enormous dragon-tail of automobiles pulled into line behind the buses. Counting the caravan of automobiles, nearly five hundred residents were now moving toward Seveso. Later, the elderly said they were left behind to guard the rooms in case of "retaliation."

The government bus drivers were perplexed. There had never been a scene like this in the routine round-trip journeys they made each day. But their puzzlement didn't last long. Before they were halfway to Seveso, each driver was approached by a quiet but determined delegation of passengers. "You are to drive on the autostrada directly to the barricades at the Corso Isonzo," the drivers were told. "You are to stop there and wait for further instructions."

The drivers protested. They were told there was no choice. They balked. Passengers in both buses roared in anger and fury. There was no alternative for the drivers. They were facing a caravan of five hundred passionately enraged countrymen, driven by righteous indignation.

By eight-thirty that morning, the captive driver of the first bus approached the turnoff from the autostrada to Seveso. The sign reading "Seveso" was covered, as it had been for three months, with tape and cardboard. The bus swung up the exit ramp toward the barbed wire, toward the sentry box. Two sentries stood there, rifles over their shoulders, staring in disbelief.

The two buses stopped short of the sentry box. Behind them were some eighty cars, backed up on both the exit ramp and the superhighway. The driver of the first bus was ordered to open the door, and he did so. Several men stepped out. The sentries slid their guns from their shoulders.

Then the children were let off the bus. Led by the men, they moved slowly toward the sentries. The carabinieri, stunned by the scene, hoisted their guns back onto their shoulders. There would be no challenge to the children. Then the crowd, getting out of cars and buses, approached, angry and shouting. Within moments, the telephone wire to the sentry box was cut, stopping any communication from the lonely post to headquarters in the town of Seveso. The sentries surrendered.

Swiftly, the protesters moved the barricades from the Corso Isonzo exit ramp, across the autostrada. Then the autostrada traffic was directed toward the center of Seveso, into Zone A via the east-west route, as the bus drivers led the way.

Traffic from Milan to Como was soon backed up for miles. Some of those trying to get through protested, but they were no match for the scowling protesters who remained behind to divert the traffic onto the Corso Isonzo and over to N-35. Among this diverted traffic were three Swiss vehicles. Their drivers were the focus of fierce epithets as they followed the orders of the evacuees, who had not forgotten that the poison came from a Swiss-owned company.

The buses and the caravan emerged from Zone A, and drove toward the Municipal Hall. Mayor Rocca came out to meet them, ordering the police back. "I do not want force to be used against my citizens," he said. "They are people who have already suffered too much."

The regional authorities arrived from Milan later. Health minister Rivolta pleaded with the people to return to their hotel. Professor Cavallaro told them of the grave risk they were taking in cutting through Zone A. There were clashes. There were demands. There were promises made by the officials, almost impossible to carry out, even with the desire to do so. The confrontation remained a stalemate.

It was uncertain how it started, but while these discussions were taking place, some people drove back along the Corso Isonzo to return to Zone A. When the cars stopped, the occupants burst out of them. They ran across the dioxin-ridden turf and into their homes. None wore protective clothing.

Some barricaded themselves inside their houses. One couple simply stood and looked at their house silently. Then they walked slowly around their flagstone walk to the garage. A screwdriver lay in their way, a lawn mower beside it. They had no key, but the door to the garage was open. Inside, a sack of potatoes sat on a shelf, its sprouts protruding. Nothing had changed since they left their home three months before. They turned around and went back to their car to rejoin the group in front of the Municipal Hall.

Soon dozens of families rushed into their homes. They clawed for the things they had missed most: books, pottery, clothes, toys, pictures. Some brought picnic baskets they had packed back at the hotel. They ate with hysterical gaiety in their own dining rooms. Neighbors visited one another, greeting friends as if nothing had happened during the past weeks. They embraced, tears streaming down their cheeks.

The macabre revelry went on until after dark. Then, gradually, the manic joy of entering their homes began to fade.

People straggled back to their cars and started down the long, desolate road to their hotel in Milan. Some did not give up so easily. They remained until the police arrived to gently move them out. By three the next morning, most of the group had left.

XIV

Continuously working behind the barbed wire in Zone A, the white ghosts of the masked technicians moved about in their heavy boots, poking and probing with their thick-gloved hands. For the most part, there was nothing but silence inside the contaminated zone.

But now and again the former residents would manage to slip back into their homes. One family lived in their house two and a half days before the carabinieri found them huddled in their living room, the last of their supplies running out.

Another woman pleaded with her daughter-in-law to take her back to retrieve a precious possession. Against her better judgment the daughter-in-law did so, waiting nervously outside the house. After futile attempts to get the older woman to come out, the daughter-in-law called, "You must come. The carabinieri are coming down the road!"

It was a lie, and the mother-in-law did not respond. The younger woman went into the house and found her sitting in the living room, clutching an old battered pocketbook, tears streaming down her face. "I'll go now," she sobbed. "I've found what I wanted."

Still another woman crept back into her home under the

cover of darkness. Each night she would slip out through a break in the barbed wire she had found, far from the sentries, and would return with food and wine, then repeat the process the next night. Her husband stole through the same barrier-break to plead with her to come out. She took a broom to him and beat him out of the house each time he tried. For over nine weeks she escaped the attention of the carabinieri and eluded her husband, until at last she gave in.

As 1977 began, the despair increased. In January, six babies were born to mothers from the contaminated zones. At first there appeared to be no problem. Then one of the six was rushed to the hospital in Milan. Surgery revealed a badly malformed intestinal system. The baby's future was uncertain.

By February, two more babies were born, one with a cleft palate, the other with a malformed large intestine. The link to dioxin was still not clear. Medical authorities were suspected of keeping the situation in low key to prevent panic. In fact, however, no doctor could really speak with authority. It had taken over four years and thousands of analyses to demonstrate that thalidomide caused birth defects. Statistics were useless at this point.

Out of the steady stream of press reports on reclamation, or the lack of it; on demands for compensation and redress; on the efforts, or lack of them, on the part of the authorities and technicians, the sudden, stinging news was announced: Genoveffa Senno, age fifty-six, died of cancer of the pancreas. She had lived with her family in Zone A, near the factory, near the Zorzis and the Cassios. Her grandchildren were two of the worst hit with chloracne, so riddled with pustules and comedones that their faces and heads were completely helmeted in grotesque white-plastic masks. Would Genoveffa's autopsy reveal the presence of dioxin? And if so, would it be announced or covered up?

February brought other ominous signs. Dozens of people from the contaminated zones began complaining of fogging of vision, of the inability to stand light, and of seeing double—all possible signs of dioxin poisoning. Liver complaints, dream disorders, cardiac disturbances, gross irritability, and loss of libido—other possible indications of dioxin poisoning—were added to these.

Then the chloracne erupted with a vengeance—130 confirmed cases—not only among the children from the contaminated zones, but among those who lived outside it. In Seveso, 200 pupils were stricken. In neighboring Meda, 110. In Cesano Maderno, 69. Counting other outlying communities, the original handful of 35 cases had risen now to over 400.

Every school in Seveso was ordered closed for a thorough cleaning. Dioxin deposits were found in two locations. These buildings had once been thought to be free of the chemical. In fact, one school had been used as the principal registration and mass testing base for the emergency program. But all through the first few weeks after the accident the gymnasium had been used by the reclamation workers and veterinarians, tramping in from the fields with dioxin on their heavy rubber boots. Students living close to zones A and B were suspected of carrying dioxin in on the soles of their shoes. The fear of the spreading of the poison increasingly gripped the minds of the people.

Reclamation workers were fighting an uncertain battle even in trying to contain the poison. Fearful about their own safety, they went on strike, demanding health guarantees. The workers were somehow placated, but in early March a tractor operator was rushed to the hospital with an acute liver disturbance. Test results were not announced. The authorities were again accused of a cover-up.

Meanwhile, the fear of the spread of the poison expanded

78

to the people of Milan. Nova Milanaise is a suburb just outside the boundaries of Milan; it had never been considered contaminated. Yet a test in a vegetable market there revealed nearly six hundred times the allowable limit of dioxin in the cauliflower. In Milan itself, the burning of animal carcasses used for laboratory tests spread dioxin over the city streets and onto a nearby fruit and vegetable market supplying Milan.

The secrecy, the lack of definite news, continued to frustrate the people. The authorities, caught between the desire to be frank and the need to prevent panic, were equally frustrated.

In April, as spring came to the Lombardy region, and the mulberry and fig trees began to bud, the joy of the season was hardly evident around Seveso. The Zorzis and the Cassios had found small apartments in Cesano Maderno, in sadness and with little hope of ever returning to the homes their hands had built.

The family of Genoveffa Senno pressed in anguish for the results of the autopsy of their grandmother, who had died the month before. There was still no word. The people throughout the region anxiously scanned the newspapers each day, wondering what would be announced next.

In Seveso, "the town where all life seemed to live in harmony with its surroundings," there was the certain knowledge that nothing of certainty would be revealed for many years. The people knew well the strange blight, the mysterious maladies, the cattle and sheep that had sickened and died, the shadow of death, and the white granular powder that no longer showed patches on the roofs. But it was there, silent and invisible.

It was there beyond Seveso, too. Quite suddenly, in mid-April, intense concentrations of dioxin were discovered in fifteen Zone B factories in Cesano Maderno. The plants were immediately closed for up to ten days of detoxification. And in

May, heavy concentrations of dioxin were found in Meda, north of the zones that had been marked as contaminated. Orders were immediately given to decontaminate the area. The specter was moving on, and all attempts to contain it seemed to have failed. And as if to punctuate this, the guarded news about the death of Genoveffa Senno had now leaked out: the autopsy proved dioxin was definitely in the liver, the first such confirmation to be announced.

There were rumors of others. Authorities would not confirm any of them. But slowly and silently the dioxin was spreading. Tests showed the concentrations of dioxin to be shifting and unpredictable, creeping downward and outward, with no certain boundaries that could contain the glacierlike movement of the poison.

By May, a silent spring had come to Seveso and no one knew how long the silence might last.

AFTERWORD

THE DAYS I spent in Seveso were long and sad. My wife, Elizabeth, who worked with me on the research for the story, found it hard to erase the scene from her mind, as I did. When we left Italy and were driving back to Switzerland, the Alps seemed to have lost their grandeur, and the green Swiss meadows their charm.

We were on our way to interview Adolf Jann, chairman of the board of Hoffmann–La Roche, and Guy Waldvogel, president of Givaudan. I had only a dim hope of getting at the roots of the disaster. The corporate responsibilities here were so massive that it was not likely they would be discussed freely.

The symptoms of the chemical plague across the world had been growing over the years, and it would seem obvious that they demanded exacting attention on the part of the chemical industry. More specifically, the history of dioxin had been flashing bright warning lights more clearly than almost any other persistent, poisonous, and insidious chemical. How could a company associated with one of the largest pharmaceutical manufacturers in the world create a condition under which such an accident could happen?

Anyone technically connected with the manufacture of any of the chlorophenate compounds would have had to have

81

been aware of the history of the two principal products derived from trichlorophenol (TCP)—the herbicide 2,4,5-T (known in the Vietnam defoliation program as "Orange"), and hexachlorophene (known by the consumer as a misguided sales hook for toiletries and deodorant soaps).

The record over the years of both these end products had been alarming, even though they contain only trace quantities of dioxin.

Take the care of a certain registered nurse who lived in Bethesda, Maryland. One day, while her three-year-old daughter was playing in the yard, a next-door neighbor was working with the weed-killer 2,4,5-T. A large cloud of the herbicide drifted over the mother's yard and engulfed her child, who for months afterward suffered from agonizing physical and mental illness. Her doctors have predicted that one more exposure to the herbicide could be fatal.

Such a prognosis is not difficult to understand in dealing with 2,4,5-T. More than 50,000 tons of it were poured on Vietnam in the abortive attempt to defoliate the jungles, and the scars remain today over four million acres. No scientist can predict when, if ever, the devasted land and the disrupted ecology of the area will recover from the dousing. Meanwhile, only incomplete results are in on the number of civilians who were killed, starved, or made to bear deformed children as a result of the chemical onslaught. Cancer of the liver has burgeoned to epidemic proportions, according to recent medical studies; many mothers have given birth to deformed children; some civilians died when they were accidentally sprayed directly.

The massive destruction was not limited to Vietnam. The impact of this persistent chemical has been felt all across the United States. Spray from crop-dusting has been known to have drifted as far as a hundred miles from the area for which it was intended. One dust storm carried 2,4,5-T all the way from South Texas to Cincinnati, Ohio.

The eventual ban on the use of the product came slowly in the United States, although the results of animal studies over a decade had clearly shown it to be teratogenic—with the capacity to cause birth defects. The tests proved that 2,4,5-T was one of the most teratogenic chemicals known, even though its dioxin content is minuscule compared to what fell on Seveso. One hundred percent of the animal litters born had at least one abnormality. Up to 70 percent of the offspring showed abnormalities including lack of heads, lack of eyes, faulty eyes, cystic kidneys, cleft palates, and enlarged livers. Dr. Jacqueline Verrett of the Food and Drug Administration sums up the results of her tests succinctly: "You would have to say this material is one hundred thousand to one million times more potent than thalidomide in the species I've examined."

It might be said that all this is irrelevant because plants like the one at Seveso are not, in fact, in the business of manufacturing dioxin. Dioxin is never produced intentionally in the manufacture of the weed-killer. It is an unwanted trace element. Yet it can't be completely eliminated in the production process.

As to hexachlorophene, even purified, it contains trace amounts of dioxin. Yet hexachlorophene was once proudly hailed as an ingredient of Dial soap, pHisoHex, Johnson & Johnson First Aid Cream, dozens of feminine sprays, cosmetics, toothpastes, mouth washes, and many other products used almost daily by the American population. Very few consumers were aware that hexachlorophene is a first cousin to the deadly Vietnam defoliant 2,4,5-T. If they had been, it is doubtful whether the advertisers would have proclaimed it so boldly.

Up until recently, it was hard to find an over-the-counter

cosmetic or cream or spray that didn't feature the words: CONTAINS HEXACHLOROPHENE. PHisoHex, the liquid soap that everyone thought to be the savior of teenagers with acne and a purifier of babies' bodies, boasted a 3 percent content of hexachlorophene. Dial soap contained 1 percent.

Years before the FDA finally and almost reluctantly removed hexachlorophene from the market, a strange series of events began to happen across the country. They were scattered. They attracted little notice.

In 1959, a newborn infant was brought home from the hospital. After his bath his mother treated him to a massage with rich pHisoHex lotion. Four days later, sores began showing up on his body. They grew worse and ugly, covering his arms, legs, and face. Then the baby began twitching uncontrollably. The twitching continued over several days and became more pronounced. Suddenly, it turned into full-scale convulsions.

He was rushed to the hospital on the eleventh day, his skin scarlet, his arms and legs jerking violently. His eyes were moving wildly in their sockets, a condition known as roving nystagmus. His face was still twitching and the left side was paralyzed.

Immediately it became apparent that the mother had failed to notice the mild warning in small print to rinse the pHisoHex off with water after applying it. The use of the soap was stopped immediately, and the baby gradually began to improve. All these symptoms turned out to be the direct result of absorption of the hexachlorophene through the normal, healthy skin of the infant.

This incident was buried in the annals of medicine, but in 1961 a scientific team headed by Dr. V. D. Newcomer studied a group of patients who had developed an ugly melanosis of the face called chloasma. The source was traced to the various cosmetics the patients were using. In turn, the only thing common in these cosmetics was hexachlorophene. In the same year another team of doctors treated twelve cases of "severe, painful dermatitis" of the scrotum, each of which was traced to the use of pHisoHex.

In 1963 a six-year-old retarded child drank about two shot glasses full of pHisoHex and was dead nine hours later. One very real problem with pHisoHex is that it looked very much like milk of magnesia and had often been taken by mistake.

In 1966, still long before any of this sort of news was being made public, a seventeen-day-old infant was accidentally given pHisoHex orally. Within four days the baby's skull began bulging strangely. His extremities went into spasms, and his face began twitching. His head thrashed back and forth, and he continuously sucked his lips in a grotesque, unnatural way. He eventually recovered, but only after weeks of uncertainty.

A year later, a group of scientists headed by Dr. F. E. Carroll reported in a leading obstetrics journal that hexachlorophene was easily absorbed through intact skin. Later that same year, two other scientists found that it remained on the skin after bathing, even when only a very weak solution had been used. Additional rinsing with water and sponging with alcohol still left a residue on the skin. In other words, if you had bathed with an antiperspirant soap containing hexachlorophene, it would not disappear with the bath water.

Adverse reports continued to mount up. In 1968 a scientist named D. L. Larson reported in the *American Hospitals Association Journal* that a 3 percent solution of hexachlorophene —the amount found in pHisoHex—passed easily through burn wounds into the bloodstream, and severe convulsions resulted in both children and adults. Other symptoms included stupor, coma, confusion, muscle-twitching, and cerebral edema.

Backing this up was a series of tests conducted at the Center for Disease Control (CDC) in Atlanta. Dr. Renate Kimbrough of the CDC found that the chemical produced definite brain damage to rats at very low feeding levels. In addition, its use in burn cases had demonstrated the same sort of damage to the human brain, as the burn victims went into convulsions and some died as a result. Further studies showed that hexachlorophene, as well as its defoliant cousin 2,4,5-T, was causing monstrous abnormalities in animal offspring.

By the 1970's, new medical reports were stacking up high at the CDC, but had not yet been published. Many of them showed that the larger the area of skin exposed, the greater the absorption of the chemical into the bloodstream. Yet babies were still being bathed from head to foot in pHisoHex, with its powerful 3 percent solution, and Dial soap, with its 1 percent hexachlorophene, was used without question in millions of American homes. Random samples of people using a soap, mouthwash, or cosmetic containing hexachlorophene began to show deposits of the chemical in the bloodstream and fatty tissues after just three weeks of normal use. One FDA scientist, acting as a volunteer guinea pig using Dial soap for some of the tests, said, "I have to admit I feel somewhat uncomfortable about all this."

Nevertheless, advertising of the virtues of hexachlorophene continued. The over-the-counter drugs and cosmetics containing the substance were flourishing. Vaginal sprays with hexachlorophene began flooding the market, from Johnson & Johnson's Naturally Feminine to Alberto-Culver's FDS spray. And when on April 1, 1971, the FDA finally got around to issuing a press release advising that any product containing hexachlorophene should be used with extreme caution, the manufacturers rose up in arms.

In spite of hexachlorophene's shoddy track record, no ban was imposed by the FDA. Despite the 1971 warning, its use continued to grow. The major supplier of hexachlorophene was Givaudan. Even as the reports of hexachlorophene's toxicity piled up like snow drifts on the desks of FDA and CDC officials, Givaudan was extolling the virtues of its product with full-page ads in the trade journals for the benefit of cosmetic and toiletry manufacturers. The advertisements were not immodest in their claims for Givaudin's own version of the chemical, which it called G-11:

> When you buy G-11 [Hexachlorophene U.S.P.], there's no escaping this fact: You know well what you're getting has been proven effective in thousands of tests around the world.... G-11 has been proven safe in every conceivable product, from soaps to shoe linings. Antiseptics to toothpastes. Through over 20 years of actual use, G-11 has been proven a moving force in the market place. The public knows it. Trusts it. Likes it. . . .

It took until January 6, 1972, for the FDA finally to announce that restrictions would be imposed on the use of the chemical. The FDA stated that "recent studies had raised questions about the use of hexachlorophene." No mention was made of the long series of medical reports over the previous decade that had made it obvious that hexachlorophene was not to be toyed with. Even dermatologists seemed to be surprised by the new restrictions, many having habitually prescribed pHisoHex for acne in teenagers, for bathing babies and other vulnerable patients.

Eventually, manufacturers of over-the-counter toiletries in the United States were forced to eliminate all hexachlorophene from their formulas. But the product continued to be made by Givaudan, both for distribution in other countries and for surgical soaps in the United States.

The ultimate horror story on the product was the tragedy in France shortly after the FDA ban in the United States. A

87

hospital attendant was preparing baby powder in a nursery ward one day, mixing the prescribed amount of hexachlorophene into the inert talc of the regular hospital supply. This was done as a normal part of the hospital routine. What the attendant didn't know was the talc had already been prepared the night before. Instead of the usual percentage, the talc now contained double the prescribed amount.

The babies were bathed as usual and powdered with the talc. Soon the convulsions and the twitching began. The condition of the infants grew rapidly worse, but there was nothing that the doctors could do. Over thirty babies simply died in their cribs.

The Seveso disaster showed clearly that even limited production of either 2,4,5-T or hexachlorophene via the TCP method could be deadly. The partial FDA ban does not stop this production, nor does it prevent the production of other equally dangerous and persistent chemicals that are contributing to the growing world-wide chemical plague.

All the ugly past history of TCP and its capacity for producing dioxin was *common* knowledge in the industry when the Icmesa plant was converted for TCP production. There was nothing secret about it. As described earlier, there had been several previous industrial TCP accidents, and though they were confined to factories, they showed clearly that high temperatures produced large quantities of dioxin, and that once dioxin penetrated deeply into a surface, it was almost indestructible. These experiences were fully described in both industrial and scientific literature.

The questions for my interview with Jann and Waldvogel of Hoffmann-La Roche were obvious; the chances of getting full and complete answers were slim.

*　　　*　　　*

A conglomeration of white plaster and varied architecture, the Hoffmann–La Roche plant stretches for acres along a nondescript street in Basel.

My wife and I were led into Adolf Jann's imposing office, from which he directs the fortunes of this multimillion-dollar organization.

A heavy-set man with a florid face, Jann greeted us cordially, offering us tea and cookies as we sat on the lounge opposite his desk. My feelings were mixed. I was steeped with the misery of Seveso and filled with anger that this company could have permitted a thing like this to happen, however accidentally. Guy Waldvogel arrived a few minutes later. A slim, youthful executive, he spoke with more composure and self-assurance than, to my mind, the situation warranted. I wanted him and Jann to feel first hand, as I had, the distress of the people of Seveso, to understand the human part of the equation which overbalanced everything. There was no question that both of these men were deeply aware of the seriousness of the situation, and they admitted their responsibilities frankly. Though they made no attempt to mask the problems, it was obvious that they tried to minimize them.

"Dioxin is a chemical we don't want to produce," Jann said. "It appears as a by-product in very small quantities in TCP. But TCP cannot be replaced by any other chemical in making hexachlorophene for very much needed surgical soaps. If we can find something to eliminate the use of TCP, we will certainly replace it. But in the meanwhile doctors and hospitals require it."

I pointed out that the United States had banned hexachlorophene for everything except professional use because of the bad track record the chemical had had over the years.

"We do not feel it is dangerous in very limited use," Jann said. "No other chemical is as effective in destroying staphylococcus."

This was correct, but I couldn't help reflecting on the former Givaudan advertising in the cosmetic and toiletries trade journals, extolling the virtues of hexachlorophene for everything from shoe linings to toothpaste.

Neither Jann nor Waldvogel were able to give a satisfactory answer as to why, after the Seveso accident, the specific warning about the dangers of dioxin came so late. Even a provisional warning would have alerted parents to keep their children from playing in the fields and romping on the playgrounds for over a week.

"When this unfortunate accident happened, our reaction was very fast. First of all, we had to find out what it was that escaped," Jann said. "We found out after three or four days. It happened on a Saturday morning. There was no full staff working. On Monday we received the samples, and we needed time to analyze them. Then we found out and gave the information out. But in the meantime we instructed the people in Italy not to eat vegetables or fruits and other things in the garden, because we didn't know what it was."

Waldvogel cut in to say: "There were samples taken immediately in the neighborhood of the reactor. There we took samples of dust to see whether dioxin could have been formed, because we knew from past industrial accidents that there had been contamination inside the factory. These samples were ready on Sunday. The reason why we could rapidly tackle the job is that dioxin control is part of the manufacturing process. But the testing is still a relatively complicated procedure, requiring at least two or three days. When we got the first results after that time lapse, we knew there was dioxin in the reactor area. Then we went and took samples from the outside. We proceeded to take samples farther and farther away from the factory—partly grass and partly wipe tests."

I tried to put myself into Waldvogel's place. The problem was that the overwhelming facts in the case almost nullified any explanation. That such a disaster had happened transcended words. The responsibilities of the chemical industry are awesome and are automatically built into their operation, just as in the nuclear energy field. There was no personal villainy involved, but certainly their experience and planning must have taken into account the question of whether such an enterprise should ever have been started at all. How did the corporation decide that the risks they could foresee were worth the benefits?

I asked the two executives more specific questions: "What is the explanation for the fact that there wasn't some kind of an automatic shut-off or a dump tank or a containment vessel to prevent this from happening?"

"This has been raised several times," Waldfogel said. "It has been oversimplified by suggesting that a tube or a vessel of some kind could have prevented the accident." He repeated his statement that he could not comment on this, however, because of the pending legal proceedings.

I had guessed the futility of asking the question, but it had to be asked. I went to another: "Would the same type of manufacturing method be allowed in Switzerland as it was in Italy?"

"Yes," Waldvogel said. "The drawings were not specifically made for Icmesa. The same drawings, the same sort of plant would have been submitted for approval in Switzerland or any other country. Whether one of these countries would have commented and asked us for some additional safety measures,

91

we don't know, we can't tell. We had a cooling system—that is why I speak of technical and human error. We had a double cooling system, and we know that there was human error in making use of it."

"Except," I said, "I talked to the worker who had to go and turn on the cooling system by hand valve. The workers claim that it should have been an automatic system."

"Well, now we are getting into legal matters," Walvogel said.

I went on with the questioning: "Wasn't it reasonable for you to assume immediately that dioxin was very probably released at the first moment, simply because it is well known that the formation of dioxin is an integral part of the manufacturing process of TCP?" I asked.

"Well," Waldvogel answered, "it is not part, it's like any chemical reaction that has only trace products when you go into purification. It may have been a certain error in appreciation on the spot after it happened. The question remains: Were we not forced to suspect that dioxin had been formed in larger than trace quantities? We only know that the temperature went to about 170 degrees. We don't know how far above that it went."

I asked: "Why wasn't any warning given to keep the kids out of the grass and the swimming pools—even with TCP alone, without the additional hazard of dioxin?"

"Well, we did say right away that something had escaped, and that they shouldn't eat from their vegetable garden, and that TCP did produce burns and other difficulties," he answered.

The interview went on for over an hour. Not unexpectedly, it did not untangle the details of this disaster. It will probably take years before that happens—if it ever does.

It is not only dioxin in Seveso and not only the continued manufacture of TCP which is a problem. Even the present supplies of 2,4,5-T are a danger. In early December 1976, five months after the Seveso tragedy, a dozen stainless-steel cylinders of 2,4,5-T, the product designated "Orange" by the Air Force, were discovered in a warehouse near Arlington, Oregon. What was startling was that the containers had been assigned to the shed by a junior engineer, who merely saw that they were padlocked inside, with a lone watchman to keep an eye on them.

When Oregon's Senator Mark Hatfield learned about the incident, he promptly demanded their removal and quoted the opinion of several scientists that the contents were "one hundred times more lethal than nerve gas."

His protests were so vociferous that the cylinders were flown by special plane to Johnston Island in the South Pacific.

In April 1977, ten months after Seveso, the Air Force revealed that another batch of the defoliant was stored outdoors in oil drums at the naval base at Gulfport, Mississippi. So difficult was the problem of destroying the chemical safely that the Air Force had been pondering it since 1970. Finally, in May 1977 a decision was made: the supply would be placed in railroad tank cars and transferred to the Dutch-owned tanker *Vulcanus*. The ship would then carry the chemical to a point in the Pacific 1,000 miles west of Hawaii, where it would be burned in incinerators at a temperature of over 2600 degrees F.

Tests will be made on the effect on marine life in the area. If they are satisfactory, the supply of the chemical on Johnston Island will be subjected to the same treatment. There are also plans to use this process in Seveso, but the funeral pyre there will be in the center of one of the largest industrial areas in

93

Italy, where far more than marine life will be jeopardized if the incinerator fails to do the job.

Dioxin is only one of the many toxic chemicals which threaten both the people and the environment. Over the years we have been quietly building Sevesos all across the United States. Several incidents have recently made major headlines. The Kepone tragedy in Virginia, the PBB contamination in Michigan, and the growing PCB contamination have not been as sudden and dramatic as Seveso, but they are dangerous themselves and they foreshadow future dangers as new chemicals come onto the market.

In Hopewell, Virginia, in 1976, when the workers at the Life Science Products Company began showing strange symptoms, including tremors, liver malfunction, and decreased sperm production, it was finally recognized that they were suffering from Kepone poisoning. Kepone, a white powder used to kill ants and roaches, is another almost indestructible compound. Careless disposal of it can create widespread contamination, as with dioxin, yet it was discovered that the company, under contract to Allied Chemical, had unceremoniously dumped huge quantities of Kepone waste into the James River. There it created one and a half million gallons of a highly toxic sludge. The environmental and population effects of this river contamination have not yet been evaluated.

Allied Chemical and Life Science were at first fined a total of seventeen million dollars, but the fine was later reduced by eight million in order that Allied might endow an environmental foundation. Despite the fine, however, how to destroy the waste remains a problem. Experiments to get rid of it have been carried out and have been considered successful, but to burn an 87-pound sample of Kepone cost $351,674—a total of over four thousand dollars a pound. One and a half million gallons of the contaminated sludge await processing.

The insidious character of the chemical plague is nowhere more evident than in the PBB invasion of the entire state of Michigan. PBB is a toxic chemical closely related to dioxin. The Michigan incident began when about two thousand pounds of Firemaster, a fire retardant with a PBB base, was mixed with a cattle feed called Nutrimaster in 1973. This accident took place because PBB was packaged in the same type of brown bag that was used for magnesium oxide, the usual additive for the feed. Very quickly after the Nutrimaster was distributed, contaminated meat, milk and eggs appeared throughout Michigan.

Soon mysterious symptoms began. Doctors were baffled by the number of patients complaining of nervous disorders, loss of memory, weakness, headaches, lack of sleep and loss of coordination. Unusual painful and swollen joints among young adults in their twenties and thirties were alarming. What's more, the victims in this age group suddenly aged neurologically. One young man had to quit his job because he couldn't remember what exit to take from the thruway. The poison penetrated into mother's milk to such an extent that Dr. Irving J. Selikoff, head of a team of researchers who conducted a health survey in November 1976 among 1,000 Michigan farm residents, said: "Until more is known, I'd recommend against breast-feeding." Dr. Thomas H. Corbett, a clinical investigator from Ann Arbor whose animal tests showed that the PBB's were both teratogenic and carcinogenic, went further: "I sure as hell don't want my kids drinking that stuff. This is not just one of the worst environmental disasters ever to hit Michigan. It may prove to be one of the worst in history."

No one knows today what the eventual fate of the Michigan victims will be. Twelve million pounds of PBB compounds have been produced throughout the country. "Ultimately, I'm concerned with the whole country's exposure," Dr. Selikoff said. "The chemical may have leached back into the environment."

Meanwhile, PCB's (polychlorinated biphenyls), chemicals closely related to PBB's, seem equally threatening. PCB's have shown up in mother's milk in Sweden, Germany, Japan, Canada, and the United States. Pregnant women and nursing mothers are now being studied for the presence of PCB's, Mirex, and Kepone in Ontario, where contaminated fish are suspected as the source.

Unfortunately there are so many dangerous chemicals that there is the problem of public apathy setting in. Hexachlorobenzene (HCB), another member of the family, has been spread by airborne industrial emissions and dumpings, leaching into the ground and thus entering the food chain. The Environmental Protection Agency is considering canceling the production of this and sixty-four other similar compounds, but action has not yet been taken.

In Louisville, Kentucky, six tons of a chemical called HEX, short for hexachlorocyclopentadiene, were suddenly discovered in the sewage system, forcing the city to close it. A minimum of nearly $9 million was needed for cleaning up the system, but in the meantime the contaminated sludge was flowing down the Ohio River to threaten the towns of Paducah, Uniontown, Henderson and Morganfield in Kentucky, and Evansville, Indiana.

The catalog continues. Vinyl chloride, a basic ingredient in plastics manufacture, threatens workers with liver cancer. Mercury continues to be a threat, not only from fish contaminated with it, but also in the vapor created in dental laboratories, where personnel who work with crowns and fillings face injury to the nervous system, red blood cells, liver and kidneys. In Pennsylvania, a stream was found to be con-

taminated with both Kepone and Mirex. Fishermen have
been warned not to eat any of the older trout, those over
twelve inches long. Eleven-inch trout are apparently not
under suspicion.

Even cut flowers have become dangerous. According to the
New York Times, flowers imported from South America have
brought with them such high concentrations of pesticides that
workers handling them in the United States have complained
of double and blurred vision, headaches, and muscle weakness. The polystyrene cup so popular in the coffee break is now
under investigation as possibly leaching dangerously into the
coffee itself.

For years the chemical industry has been vociferously op-
posing regulation of its 30,000 different chemicals. Partly because of this opposition, there were five years of hearings
before the Toxic Substances Control Act was finally put into
law in October 1976. Most nonindustry specialists, however,
consider this law only a small step in the right direction. The
EPA decides what chemicals are to be tested. The tests are
loosely defined, and preparations for them will probably take
until late 1978, according to EPA officials. PCB's, for example,
will not be totally banned until January 1, 1979.

Regulation and testing of chemicals already developed are
obviously slow in coming. But a problem also arises with sub-
stitutes for these chemicals. As Dr. Selikoff states: "Immediately the question comes up—how do you know the substitute
is not *more* dangerous?"

There is little doubt that the potential for massive catastrophe is growing. Weak legislation is being outraced by uncon-
trolled production, where profit takes the front seat instead of
safety. The risks and benefits are not being weighed and bal-
anced. Society at large is not being made part of the decision-

making process, yet every decision affects not only local communities but the world. The effects of industrial production are no longer localized. Technical decisions have become social decisions. When chemicals like dioxin, PCB's, and PBB's can turn mother's milk into a poison, something is wrong. When a single nuclear power plant has the potential of contaminating an area the size of Pennsylvania, creating a Seveso-like condition over 45,000 square miles for a population of over 10 million, something is wrong. When you can't buy an insurance policy anywhere to protect against such a disaster, something is wrong.

What happened in Seveso was a tragic dramatization of what we all risk without better control. Only a personal visit to Seveso can fully reveal the fear, the confusion, the frustration of everyone in the area. The people have nowhere to turn. They grasp at straws of hope. Whether I talked with scientists, doctors, lawyers, officials, or townspeople, the sense of loneliness and separation from the rest of the world was evident always.

I was in Paris in mid-August 1976 when I was assigned by the *Reader's Digest* to investigate the story. It was hot, and Europe was in the middle of one of its worst droughts. Bottled water was more expensive than many of the French wines and it was hard to come by. The coverage of the Seveso explosion in the French press was sparse, but there was enough to suggest the magnitude of the tragedy. Already, experts from England and the United States were being flown to Milan to attempt to see what could be done. One of the companies involved was the British chemical decontamination firm of Cremer and Warner. I decided to phone them before driving down to Italy, to try to get some kind of overview.

Alex Rice, one of the firm's engineering specialists, was cordial and communicative. He had just returned from Seveso and was in the process of working out detailed plans for combating the spread of the poison, but because the zones had already been shut off by the carabinieri and army units, he was having difficulty in appraising the magnitude of the job. The provincial government of Lombardy was touchy, he said, because it did not want Givaudan or Hoffmann–La Roche to appraise the devastation. They felt there would be too much likelihood that these interested parties would minimize the dangers. Cremer and Warner would therefore report their appraisal directly to the government.

Was he afraid of the personal risks of tackling such a precarious job? Mr. Rice was guarded in his reply: "I would think that one knows the risks and applies precautions, and then hopes for adequate protection," he said. "There is no point in unnecessary risks. But we have to think of the people involved. They need help desperately. We are going to try to spot the hazards, and advise them in precautions and safeguards. There are no such things as certainties in this life."

To Mr. Rice, the problem boiled down to logistics, even in the way he would be taking off his protective clothing. "This will be a dangerous step," he said. "We can't be too careful. Actually, we will be using the same techniques as in cleaning up nuclear contamination. The whole job will be the same as the other jobs we do—we constantly have to think about the unthinkable. And I keep thinking about what would have happened if the wind would have shifted just five degrees in either direction. It would have literally hit a hundred times as many people. It's bad enough the way it is."

Mr. Rice's tone was calm and precise, but the enormity of the problem he was facing was obvious in what he said. To go to Seveso and plunge into the massive decontamination was a job far beyond the ordinary line of an engineer's duty

It was not only Mr. Rice who faced formidable problems in

handling the Seveso situation. The quiet scientific routine of Milan's Mario Negri Institute was turned upside down. "Our first problem is to establish clear methodology under crisis conditions," Professor Silvio Garattini, the director of Mario Negri, told me. "It involves analyzing vegetables, soils, and animals, because they are the key to everything, especially to delimit the territories that are uninhabitable. But dioxin requires some of the most sophisticated equipment possible. We're lucky to have as much as we have. Fortunately, we had previously tied in with the National Institutes of Health in the United States and can deal with the problem. We were already working with a computer analysis program with Stanford University, and that could be converted. So international cooperation paid off when we least expected it."

One Mario Negri scientist looked beyond his emergency work to the broad social picture involved in the growing incidents of the chemical plague. "We have to look at the benefit/risk ratio in the production of these chemicals," he said. "For instance, vinyl chloride is at the heart of plastic manufacture. It's definitely carcinogenic, there's no question about it. Yet if we tossed it aside, we might have six million unemployed on our hands. We've never before faced a world-wide problem like this in history. The scientists and the experts should not make decisions like this alone. It's a moral, ethical, and political problem. And the decisions have to be made by an informed public with steel nerves."

Like most of the other scientists at the Institute, he could hardly believe the news that so much raw dioxin had been permitted to escape from the Icmesa stack. He joined the others on the staff who asked *why* there was no dump tank or other safety device to contain such a poison.

Dr. Luciano Manara, the scientist who had phoned the news of dioxin's terrifying potency to the Seveso mayor's office, was concerned about the indifference of the public to an industrial society that was creating so much potential havoc to the environment. "The people who are traditionally responsible for the economic culture are simply not aware of the weight of science, and how it affects our lives," he said. "An accident like Seveso can happen any time, anywhere. People are paying a tremendous price for this scientific culture. Can you imagine how I felt when I had to phone the mayor with the news? It was a horrifying experience—both for him and for me."

Walking through the long corridors of Mario Negri, I found similar reactions from nearly every scientist I talked to. They still seemed stunned that such an accident could happen. At the new emergency lab miraculously set up within days, Dr. Alfred Frigiero sat by the mass spectrometer rushed down from Sweden, and speculated about the accident.

"The factory was producing trichlorophenol—TCP," he said. "There are always trace amounts of dioxin in this process. But if the temperature goes over two hundred degrees C., dioxin begins to form. This is well described in the literature. If the safety valve blew, the chemical has obviously overheated, and dioxin is released. In fact, no other product can be created except dioxin. And the higher the temperature, the greater the amount. The question then is: Why did Hoffmann–La Roche take nearly ten days to announce the fact that dioxin was released over the entire area?"

This was the big question, and it was unanswered. Practically every one I talked to brought it up, mostly in anger. The estimates of the amount of dioxin released varied. The lowest was two kilograms, which could kill a million rabbits. The highest estimate—260 pounds—could literally kill billions of rabbits. The human equation is still in doubt.

At the Istituto Ostetrico Ginecologico in Milan, Dr. Francesco d'Ambrosio was angry. "Roche knew from the samples that were taken back to Switzerland that dioxin was present. Within forty-eight hours, they knew this. They should have known it immediately, anyway. And they did not come here to tell us until at least ten days later." Then he added bitterly: "It's very strange, what happened to the cloud that came out of the stack. It seems to have been a magical cloud. It skipped and jumped, and left some houses clear of dioxin and others not." Dr. d'Ambrosio saw in these arbitrary decisions a political gambit, a reflection of the struggle between the Christian Democrat- and Communist-dominated areas. Others less politically motivated saw such actions as ineptitude or honest confusion.

"The people in the Communist area are allowed to remain living in Zone B, but they can't work in the area. I want to know the reason for this. Why can they live there, but not work there? Is there a danger? Is there not a danger? Further, nobody knows how much dioxin is in the area. Or if they do, they won't say. I feel that they are not recognizing the dying animals of this zone because they were Communist, and not Christian Democrats."

Aside from the political inferences, Dr. d'Ambrosio had little respect for the artificially drawn zones. It was impossible, he said, to maintain that Zone A was clearly divisible from the other zone or to think that dioxin would respect any barbed-wire barriers.

Like several others, Dr. d'Ambrosio suspected that there were military aspects to the situation that were not announced. Although Roche denied it strongly, there was suspicion that the production of the TCP was not merely for surgical soaps, but that its manufacture for some form of chemical warfare was involved. It was a question that also remains unanswered to this writing.

*   *   *

I have to confess I postponed going into the contaminated areas of Seveso as long as possible, gathering material in Milan to put off the day. As a devout hypochondriac, I had to build up a slow psychological adjustment. I think that for most people, the worst part of dioxin is that it is invisible. If it showed itself in drifts, like snow or sawdust, you could pick your way around it and avoid direct contact. As it is, Zone A and Zone B looked no different from the rest of the area.

I came to admire people like TV reporter Bruno Ambrosi and his assistant Piera Rolandi for their tenacity in staying on top of the story. Like nearly everyone with any interest in Seveso, both reporters were furious at Givaudin's delay in letting the people know about the possibility of dioxin contamination, and exhausted from troubling an apparently deaf heaven with their bootless cries. Ambrosi can be credited with breaking the story wide open. His background in chemistry and passion for medical stories was an immeasurable help. His call to Dr. Manara was the ignition that put the entire region into action. "Journalists, because of their profession, have a bit more fantasy, or rather imagination, than officials do," he said. "More fantasy in the sense that the imagination is free to wander beyond the bureaucratic boundaries, which are very confining. The people and the officials really didn't understand what was going on. They really weren't aware of the danger at first."

Piera Rolandi expected the story to be over in a week to ten days. Instead, she and Ambrosi were making repeated trips to Seveso, so many that the doctors insisted they have regular medical and blood tests along with the inhabitants of the area. They had been in the worst contaminated zones for ten to twelve hours a day, before these areas were closed off, and they had watched as the danger zone was enlarged almost daily. "All the people seem to tell me the same thing," Ms.

Rolandi said. "They don't want any compensation for profit. All they want is to replace what they had." It was from Rolandi that I learned of a woman of sixty-five who had saved all her money to leave Milan and retire in Seveso so that she "could breathe good fresh air."

"The spirit of the people I interviewed in northern Italy where there was an earthquake was quite different from Seveso," she said. "Although their houses had been totally destroyed, they were full of courage, doing everything they could to try to reconstruct them. Seveso is much more tragic because the people can still see their houses sitting there, apparently intact. Sitting right in front of them on the other side of the barbed wire."

The reporter paused a moment, then continued. "It is physically and psychologically different from an earthquake. The situation in the earthquake area was a natural disaster. It was not caused by man. And so the anger at Seveso is much deeper because of that. When I go to Seveso, I feel as if I am living in some kind of science fiction story. Everything seems so unreal. You are not able to touch anything. People keep asking: 'Why can't we go back?' This is the most disturbing thing in my interviews. It is very upsetting."

When you first enter the ancient town, it looks almost normal. Route 35 from Milan is a road much like any strip road found in hundreds of American cities, a suburban highway crowded with trucks, cars, and cyclists. Although it doesn't cut through Zone A, as the autostrada to the east does, most of the cars have their windows tightly rolled up, the occupants literally baking in the summer heat. The cyclists, however, are totally exposed, riding through the dust churned up by the trucks and cars. The pedestrians share the same fate.

Texaco, Total, and BP service stations are sprinkled along

the road, among modern furniture *fabbricas* and their well-kept showrooms, with signs proudly proclaiming the craftsmanship of the Brianza regions. They in turn are interspersed with billiard-flat cornfields. Occasionally a modern high-rise condominium towers over old burnt-yellow apartments and mud-brown stucco villas with red tile roofs. Numerous little *pizzerias* and *ristorantes* complete the roadside scene.

When the town line of Seveso is reached, the signs reading WARNING—ZONE OF POLLUTION begin to appear. But there are also other posters reading SEVESO—OUR LIFE CONTINUES. These go on to implore the people to pick up the pieces of their lives and to demand justice and compensation for their losses.

When you get out of your car in the town, one thing is very noticeable: you don't hear any crickets and you don't see any birds or pets. When my wife and I arrived, we went directly to the handsome new elementary school set up for the emergency program, a stone's throw from the barbed-wire border. Nearby, sentries in combat uniform stood silently beside their hastily constructed guardhouses. The parking area in front of the school was bustling and active. Mothers were leading their children in and out of the main portico for the constant medical tests that the citizenry was facing.

A classroom on the second floor was headquarters for the Icmesa union workers, who gathered in clusters awaiting the results of their tests and discussing their uncertain future. Bitterness and anger against the Swiss owners and the local Icmesa management was the keynote of their mood. The titular head of the union was a handsome man in his thirties with a thick beard and dark, penetrating eyes. He had no love for American journalists or multinational corporations. He was convinced Americans were directly or indirectly behind the production of the trichlorophenol for the manufacture of the

2,4,5-T defoliant or even for chemical warfare use of dioxin. He felt strongly that American workers should join with his union to curb the production of anything that contained even minute quantities of dioxin. "We feel sure that half of this production of TCP is being done for America," he said. "If so —what are they doing with it?"

I went next to the Seveso Municipal Hall, a musty, venerable, and imposing building. Mayor Rocca sat in his paneled office, a sad but impressive figure of a man frustrated in his desire to help his people out of their misery. Built like a heavyweight fighter, he was sensitive and literate.

"When you see the rabbits dying, the cats and dogs swaying, and the birds falling from the sky, your heart is overwhelmed with both sorrow and terror," he said. "The people, however, have accepted the tragedy with great dignity and civic pride, in spite of their occasional impulsiveness. We have the feeling now that we are sort of guinea pigs, and this we don't want. We want to remain living on our own land in spite of the dioxin. We want to work here as we always have. To have our traditions, as we always have. And to have the comprehension of the whole world."

He continued: "We have found ourselves in the middle of the hottest controversial subject in the world. This is an event that literally questions our civilization. The emergency moments are behind us, and we must now face great social and economic problems."

Just outside the Municipal Hall, across an ancient courtyard, is the tiny, cramped police station where Chief Alfonso Calo and a handful of local police handle the routine minor infractions of the law that have characterized Seveso over the years. When we arrived at the station, the chief was at work with two officers on loan from Milan, temporarily replacing Seveso police who were literally in a state of exhaustion. All three were in the process of sorting out index cards filled out by the families evacuated from Zone A. The cards listed how much property each family owned, how much land was cultivated with fruits and vegetables, how many sheep, cattle, and rabbits had died or had to be slaughtered. It was a tedious, confusing job, which reflected the endless minutiae the disaster had brought with it.

At the gymnasium in the basement of the elementary school, the collection of animals and soil samples continued. The town veterinarians were gathered by a board posted with a huge area map. They were carefully placing their stickers on the map according to the exact locations where the animals were found.

There was a warm camaraderie among these men who had been working together, with little sleep, since the cloud had come over. They talked of the early days after the disaster when they had picked up dead animals with their bare hands, unaware of the dioxin and believing that only the trichlorophenol was involved. All agreed that the devastation of the animals' viscera revealed in the first dissections of the carcasses was literally shocking.

Monsignor Guzzetti, spokesman for Archbishop Columbo of Milan, was a quiet, soft-spoken man who methodically presented the attitude of the Catholic Church on the abortion issue. This was one of the most critical questions raised by the disaster, for it struck at the core of the Church's position on the subject. The dilemma was obvious.

I was a little surprised at the signs of flexibility when Monsignor Guzzetti talked about the Church's position: "The Church has never condemned or passed any judgment on the women who have decided to follow the road to abortion. However, the Church has emphasized the importance of the life of the child even if it is deformed. It has a life and it is important."

Then he added: "The Church asks the people to stay close to the ones who are affected by this problem to try to help them emotionally and morally. To be kind and close to them."

In spite of the sympathetic stand of the archbishop he continued to insist on the mothers bearing the children, and putting them up for adoption if the parents were unable to care for a malformed offspring.

This position created major controversy and further inner conflict for the pregnant wives. As in all aspects of the catastrophe, confusion and paradox reigned. It was difficult for a woman bearing a child to tell which way to turn. Guzzetti reflected this in stating the Church's position. "The Church understands that people are in a particularly dramatic situation," he said, "in which they cannot necessarily be responsible for the decisions and acts that they take at this moment. These people are people who must be loved." He went on to point out that the Church was taking active steps in other directions, to help with housing, social adjustment, finding safe play areas for children, and instructing couples on natural contraception methods.

Everyone placed in any position of authority or policy-making became a target for the displaced people. In such a situation, nothing that was done was right. As an outsider trying to track down the complicated threads of the story, I had difficulty enough. The key figures of the drama were scattered out all over the labyrinth of Milan's streets and out into the suburbs. Merely locating them was a chore. And many of those I interviewed reflected a strange combination of an attitude of brotherhood and of criticism toward those whom they worked with. There was a great deal of in-fighting, but at the same time there was evidence of sublime cooperation.

One of the most dedicated men of authority was Lombardy's minister of health, Vittorio Rivolta. When we arrived at his office, the strain from his work and the lack of sleep over many weeks showed clearly in his face. A ruddy, handsome man in his fifties, he shared his information openly, reflecting a deep concern for the people and the countryside he obviously loved. Yet he was the focal point of fierce resentment because the problems were not being solved quickly. It was hard to accept that there were no simple solutions, that no one in authority could do anything that would erase the curse that had fallen on the town and the area.

"The mystery of the substance is incredible," Rivolta said, sitting in his office at a desk piled high with statistical reports. "The extent and danger created by this tiny molecule doesn't change the people's outlook. Nobody even imagined that a little factory could create a disaster of these dimensions. They were confronted with a catastrophe that was difficult to face because of this mysterious quality. Think of it. A substance so powerful that five micrograms in one square meter of soil creates a deadly toxicity. The instruments to detect this are few. There are only five mass spectrometers in all of Italy. A single test analysis of soil or tissues costs over four hundred dollars in America. We are doing thousands of them. This is the first case in the world of dioxin contamination over a large populated area. A poison that creates mutations, liver destruction, malformed children, neurological destruction, and cancer. Everything is insidious about it—especially its psychological effect on the people. We have to explain to them constantly what we are facing. It is hard to get this across."

It was even harder for Rivolta to explain how little could be done to remove or destroy the poison. Both he and the people were caught in the throes of a chemical plague that could have happened almost anywhere. As we drove through Milan to the Leonardo da Vinci Hotel, it was easy to understand the disquietude of everyone concerned.

The victims of Seveso were assigned to one wing of the structure, one that could compete with any of the better Hilton or Sheraton hotels, but the fact that the hotel was pleasant in atmosphere did little to mollify the hapless evacuated families. Scattered about in the lobby were a few families talking softly with each other, as children scampered around the rugs and furniture. The people were well-groomed but their body language reflected the mood of tourists facing a long-delayed flight at an airport. Some simply sat and stared ahead, waiting for something they could not define.

One middle-aged woman told me: "It took us twenty years to build our homes. We built them with our sweat. Our husbands worked hard all during the week on their regular jobs. They worked Saturdays and Sundays to build. We don't want to go to apartments they are trying to find us. We want the money to replace our homes. Hoffmann–La Roche has promised to pay us for them. The authorities promise a lot, and they do nothing. They don't want to evaluate the homes because they say they are afraid to go into the poison zone. We told them to make an evaluation from the design of the house, the blueprints. All they need to do is to pay us for what we had. We now have nothing. We are afraid that if the company pays the region and then the region pays us, a lot of money will go into their pockets."

Another woman, markedly upset, spoke up. "We are afraid to go back, too. Even if they gave our homes back to us, we couldn't sell them. The whole area, they tell us, will become a desert after it's closed off for good."

For two days we talked with other people at the da Vinci,

and the story that emerged was a tapestry of despair and longing. The sprouts of the plan to make the desperate breakthrough into Zone A were already showing then, weeks before it actually happened.

At the office of Pier Torrani, the attorney retained by the province of Lombardy, we talked at length about the unprecedented extent of the damages. It was almost impossible to estimate the losses of property, income, health, conscious pain and suffering.

Torrani, a perceptive man who looked like an English country gentleman, was determined to press a fair but firm case against Hoffmann–La Roche and its subsidiaries.

"There are so many kinds of damage involved," he said, "it's doubtful that any kind of insurance could cover them. First, there is the cost of public administration, including sanitation clean-up, housing reclamation, real estate and economic losses. As widespread as this is, it's relatively easy to figure after we get the costs assembled. Then there are the animals and livestock that died or had to be killed. That's relatively easy to compute.

"But the most important—the damages that can't be calculated for years—these are impossible to estimate at this time. The full effects on the people will not be felt for five, ten, or fifteen years—if then. And what about the indirect damage to the community—the fear and psychosis that have permeated the public? It's all staggering. Hoffmann-La Roche has made some very generous statements to the press about paying for the damage, but we haven't yet seen it in the form of a legal document."

Attorney for the defense on the scene in Milan was Alberto Visconti. He was guarded in his statements, as he obviously had to be. He had nothing to say about the amount of insur-

111

ance available. He noted that both the Swiss companies and Icmesa were doing everything they possibly could on reclamation, continuing their experiments daily, but it was too early to tell what the results might be. He pointed out that his clients were paying directly all the costs of the evacuees at the hotels where they were billeted, and that Roche had offered to pay for the destroyed crops for 1976 and 1977. He also emphasized that the conflict between the industrial group, the regional authorities, and the victims was slowing down the indemnification program. "The region just doesn't want to let go," he said.

Visconti was emphatic in denying the rumors that the Icmesa plant had been producing defoliants or selling its production for military use by NATO or the United States. "There is no such contract whatever," he said. "The product was being made only for ultimate use in surgical soaps."

I asked him the critical question: Why was the warning about dioxin delayed for so many days?

Visconti said the situation broke down into three stages. First, it was believed that only the solvent escaped. Second, it was thought that only TCP was involved. Third, it was finally realized that dioxin was specifically contained in the cloud. His comments, however, did little to explain why the presence of dioxin was not immediately suspected and announced.

We could leave Seveso physically, but we could not leave it emotionally. The memory of driving through the heart of Zone A on the autostrada was particularly sharp and poignant. The signs were still there: CONTAMINATED ZONE. ROLL UP WINDOWS. CLOSE VENTS. DO NOT STOP. DRIVE SLOW. The barbed wire stretched along both sides of the thruway, a nonpolitical Berlin Wall. In the summer sun, the heat was unbearable; the car was an oven. There was very little traffic as the cars moved slowly through the one slice of the zone open to traffic. The tape on the sign at the Corso Isonzo turn-off, covering the name SEVESO, was a silent reminder of the tragedy.

On the western side of the barbed wire, the varicolored houses stood like mausoleums, silent, shuttered, but apparently physically unscathed. The streets were empty, but in a field in the distance a group of technicians in their masks and white space-age uniforms could be seen moving awkwardly through the grass, as if they were taking the first steps on the moon. The silence and emptiness were ominous and otherworldly.

Somehow I felt Rachel Carson should have seen this place. But perhaps, as she wrote *Silent Spring*, she already had.